GILBERT WHITE was born in 1720 at Selborne in Hampshire, the parish that his life's work, *Selborne* (1789), was to immortalize. After a country childhood, during which he absorbed daily the wisdom of hanger, meadow, and stream, and school at Basingstoke, he was admitted to Oriel College, Oxford—an election that later, after ordination, prevented him from ever becoming vicar at Selborne, for the parish was in the gift of Magdalen. Entry to the world of academic scholarship became allied to a passion for field diversions; and in September 1741, during the pursuit of partridge, it was the experience of a fall of gossamer that led to what a later writer would call 'a sense sublime | Of something far more deeply interfused' and to a Damascene conversion—from a raptorial to an observing attitude to Nature.

From this time on, despite a later bid for academic preferment, White's future was determined. In 1751 he began a 'Garden Kalendar' and started the quiet elaboration and extension of his grounds that was to continue for forty years; he studied botany, completed a flora of the parish, and corresponded with naturalists of the period—notably, Thomas Pennant and Daines Barrington; he wrote papers for the Royal Society on swallows, swifts, and martins (both house and bank), advised a brother at Gibraltar on how to write natural history, and—eventually—completed his own volume. This last was his sole published book and he died a few years later, on 26 June 1793, in the home at Selborne where he had lived since 1728—and which now houses a museum in his memory.

PAUL FOSTER is Head of English at West Sussex Institute of Higher Education. He has published extensively on White, including *Gilbert White: A Scientific Biography* (London: Christopher Helm, 1988); he also edits the Otter Memorial Papers, a series of short devotional monographs which adopt a multi-disciplinary approach to local artefacts such as the Arundel Tomb, the Chichester Reliefs, and the Chichester Tapestries.

THE WORLD'S CLASSICS

GILBERT WHITE

The Natural History of Selborne

Edited with an Introduction by
PAUL FOSTER

Oxford   New York
OXFORD UNIVERSITY PRESS
1993

Oxford University Press, Walton Street, Oxford OX2 6DP

Oxford New York Toronto
Delhi Bombay Calcutta Madras Karachi
Kuala Lumpur Singapore Hong Kong Tokyo
Nairobi Dar es Salaam Cape Town
Melbourne Auckland Madrid
and associated companies in
Berlin Ibadan

Oxford is a trade mark of Oxford University Press

Editorial material © Paul Foster 1993

First published as a World's Classics paperback 1993

British Library Cataloguing in Publication Data
Data available

Library of Congress Cataloging in Publication Data
White, Gilbert, 1720-1793
The natural history of Selborne / edited with an introduction by
Paul Foster.
p.   cm.—(The World's classics)
Originally published: 1789.
Includes bibliographical references (p.   ) and index.
1. Natural history—England—Selborne.   2. Selborne (England)
I. Foster, Paul G. M.   II. Title.   III. Series.
[QH138.S4W5   1993]   508'.422'74—dc20   93-486
ISBN 0-19-282928-9

1 3 5 7 9 10 8 6 4 2

Typeset by Cambridge Composing (UK) Ltd
Printed in Great Britain by
BPCC Paperbacks Ltd
Aylesbury, Bucks

CONTENTS

Acknowledgements   vii
Introduction   ix
Note on the Text   xxv
Select Bibliography   xxviii
A Chronology of Gilbert White   xxxiii

The Invitation to Selborne   I

THE NATURAL HISTORY OF SELBORNE   5

Letters to Pennant   9

Letters to Barrington   99

White's Notes   251

Explanatory Notes   262
Biographical Notes   296
Glossary   302
Index   309

ACKNOWLEDGEMENTS

To read *Selborne* is to discover White's debt to his contemporaries and, as the work becomes increasingly literary, to the classical authors also. These last, from at least Letter 26 to Barrington, frequently provide him with epigraphs for his letters and enable him to maintain a continuity between the two worlds, the literary and the scientific, in which he moved. So, in a modest way, is it in my case, and I record here my debt not only to the many professional librarians and others who have assisted in providing the continual supply of information on which editors are dependent, but also to Sharon Shopland and to Samantha Elliott, each of whom has made a distinctive and valued contribution. For all else, I acknowledge the guide of Calliope, and dedicate this edition of *Selborne* to the memory of the late Joss Hiller and Margaret Grainger.

P. F.

Chichester

INTRODUCTION

> Ye shall describe the Land . . . and bring the Description
> to Me.
>
> (Joshua 18: 6)

GILBERT WHITE (1720–93) was born in Selborne, a quiet
country village in Hampshire; he lived in the village for most
of his life, and died there. Late in the 1780s he completed
almost twenty years' work on one book, *The Natural History and
Antiquities of Selborne* (1789). Publication of this volume, White's
sole full-length work, was arranged through his London bookseller brother, Benjamin; a review, by another brother,
Thomas, appeared in the *Gentleman's Magazine* for January and
February the same year; and in 1802 and 1813 new editions,
published by a nephew who had inherited Benjamin White's
business in books, were made available.

Set out thus, it is difficult to envisage a literary work with less
claim to fame. At a time (1990s) when our attention is presented
daily with national and international issues of the most
intractable kind, 'parochial history', to use White's own definition from his Advertisement to *Selborne* (see p. 7), is not a genre
likely to attract much notice. Further, a volume that took
twenty years to prepare and that was published, reviewed, and
republished by relatives of the author suggests a venture of such
indulgence that merely to open the pages must risk a trespass
on familial privacies. Better by far, we might decide, to leave
Selborne on the shelf, seek the nearest wildlife habitat, and study
the bees and the birds for ourselves. In doing just that, however,
we should be following precisely White's own hope. His
Advertisement defines not only an understanding of 'parochial
history' but also the purpose of *Selborne*. This purpose, in so far
as it relates to the natural history portion of the volume,[1] is

---

[1] White's complete text, with a comprehensive title-page, falls into two parts
each prefaced by its own title, vignette, and motto(es): the first part is devoted to
natural history, the second to antiquities. Subsequently, in using *Selborne* as a
term of reference, I refer normally to the natural history portion of White's volume.

twofold: firstly, to induce 'readers to pay a more ready attention to the wonders of the Creation [which, White thought, were too frequently overlooked as common occurrences', and secondly, to advance 'the enlargement of the boundaries of historical and topographical knowledge'. Such a dual purpose strikingly rebuts any charge of self-indulgence; although the observation of natural phenomena may be a matter of individual pursuit (and yield much pleasure), the aim of the pursuit is 'the enlargement of . . . knowledge'—which has to be communicated in books such as *Selborne*. And exactly there is the justification for reading it. As with so many explanations for human action, *Selborne* exists: it is a book written, as White claims, to be shared—to be read. And read it has been! In the two hundred and more years since first publication, *Selborne*, in one format or another (for the volume's capacity to be interpreted variously by editors and illustrators is a feature of White's achievement), has been published on over two hundred occasions. Durability and accessibility in such measure demand interpretation, and it is the prime purpose of this introduction to come at some explanation of why this has been, and is, so. What is it about *Selborne* that has prevented it from being left along with hundreds and thousands of other volumes about natural history to moulder quietly away on some dusty shelf, having lived for a day and been forgotten on the morrow? My own answer falls into two areas: White, and his methods of working; *Selborne* itself, and what it offers.

White, and his Methods of Working

At first sight White's life offers considerable promise of the future *Selborne*. To the advantages of a country childhood in the parish that his writing was to memorialize was added not only the influence of a paternal grandmother whose local connections were of great value (Rebecca Luckin was daughter of a yeoman farmer in a parish adjacent to Selborne), but also the importance of a classical education. This last, for what White was later to accomplish, has a more significant bearing on his development than is sometimes recognized. The future writer within the rural tradition was born into an educated and literate home, traditional in character (his paternal grand-father was vicar at Selborne for over forty-five years and his father was called to the Bar in 1713), and he himself proceeded to university. All this is a commonplace; and that when at Oxford (Oriel College) he should read contemporary literature (for example, the poets Edward Young and William Collins, and Samuel Richardson, the novelist), write verses himself in the fashionable heroic couplets of Dryden and Pope, study and translate from the classics, is also unexceptional. And again, that such orthodoxies should lead to ordination and curacies is not unusual. But for some reason, acceptance of these conventions led White to overestimate his abilities. Office as a proctor of the university and dean of college pointed (he thought) to academic preferment and in 1757 he offered himself for election to the provostship of Oriel. He was not successful and this, together with the death of his father the following year, was to prove a turning-point in his life. At first he withdrew into familial pursuits, including a tour for six months to visit relatives in Rutland; then, a few years later, he inherited Wakes, the home in Selborne that has become a mecca for White enthusiasts, and the freedom to reassess his future. It was to be given to the natural calendar and to the records that were to contribute so powerfully to the eventual *Selborne*.

Observation of the cycle of the year and interest in the vagaries of the seasons is common among country people, but in White's case several impetuses were at work. Since 1751 he had gained experience in maintaining detailed records for Wakes garden that embraced planting, tending, and harvesting; but now, early in the 1760s, was added knowledge of *Miscellaneous Tracts relating to Natural History, Husbandry, and Physick* (2nd edn., 1762) by Benjamin Stillingfleet. This volume, which incorporated a natural calendar for Uppsala and for Stratton (Norfolk) for the year 1755, fired White's enthusiasm. Although his garden records had occasionally included mention of birds and other creatures, there now came to him the realization that his propensity for keeping records for personal and familial interest could serve a larger pur-pose—the improvement of the quality of life for all people.

Such a high purpose was wholly in the spirit of the age (many famous hospitals, for example, were founded in the middle decades of the century), and articulates the humanitarian concerns of the period.

White's hope, however, derived from a longer tradition—the programme enunciated by Francis Bacon (d. 1626) of forming a comprehensive collection of data about the natural world which, once assembled, analysed, and interpreted, would establish the laws of the natural world, both physical and biological. For the biological world the serious beginnings of the programme had awaited the attentions of Carl Linnaeus. By means of his own passion, the sure grasp of a systematic approach (however artificial its principles), and specimens collected by himself or supplied by many disciples travelling the world, this Swedish naturalist prepared listings of species world-wide. These were published in *Species Plantarum* (1753) and *Systema Naturae* (10th edn., 1758), and were to become the starting-points for all subsequent naming. But Stillingfleet, as well as becoming an advocate of the Linnaean programme (he is attributed with guiding William Hudson in preparation of *Flora Anglica* (1762), the first British flora on Linnaean principles), added a classical dimension. Appealing to Hesiod, he suggested that a correlation might exist between the activities of nature (the migration of birds, and the flowering of plants, for example) and the needs of the husbandman; in other words, that the times of ploughing and sowing, or the very best moment to harvest a crop, might be evident from observation of an appropriate prognostic. History showed that the lot of mankind was materially affected by the price of staples. A poor harvest, whatever the cause—the loss of seed-time or the destruction of a crop by inclement weather (phenomena that White often records in his Journals)—was the most important factor contributing to the anxieties of the nation; and if, by simple observation and record, some stability of production could be achieved, a worthwhile advance in the management of human affairs would be gained.

For White, with his close knowledge of the natural scene, his broad human sympathy (throughout his life he attended the poor, the ill, and the bereaved, and maintained the charitable dispositions of his grandfather), and his resolute belief in the beneficence of Providence, Stillingfleet's ideas were compelling. Not only would they provide, in an active, rational fashion, a mode to further the demonstration of 'the wisdom of God manifested in the works of creation' (a cause much in tune with White's heart, albeit announced by John Ray in 1691 in a volume of that name), but their successful exploration might answer the mental and spiritual disturbance that he had experienced twenty years previously. This experience, in autumn 1741, had occurred during a foray to shoot partridge. The details, of how his dogs were incommoded, and of a subsequent shower of gossamer, may be read in *Selborne* (see Letter 23 to Barrington), but the effect, of White 'musing in [his] mind the oddness of the occurrence', lived with him for the remainder of his life. In one sense, the leading features of this experience, of the kind that in 1805 Wordsworth was to characterize as 'spots of time' (*Prelude*, xi. 258), are unremarkable: after all, it is not uncommon for young men to engage in field-sports (and as a young man White pursued them actively) and to be hindered in the execution; but in another sense the whole experience, located in particular on the mode of hindrance, was compelling. To fail in the hunt through one's own incompetence, through the perversity of one's dogs, or through the subtleties of the game, were acceptable aspects of the chase, but, as in White's case, to be forced from the field by gossamer, by the slightest, lightest, seemingly most delicate of phenomena, was a mystery beyond comprehension. And if now, twenty or so years later, Stillingfleet, with his proposal to record the comings and goings of birds and other creatures alongside the flowering of plants, offered a pathway towards uncovering the mystery and demonstrating the inherent beneficence of the universe (both for himself and for the wider world), then it behoved him to pursue it.

Such a high purpose would demand rigorous preparation. In a measure this was already accomplished. White's classical education had provided him with fluent Latin; hence, as well as learning about natural phenomena from talk with ordinary people, he was able to read the main scientific texts of the time. As a consequence, when in 1765 he initiated a close study of the plants in his parish, he was able to use as his main text Hudson's *Flora Anglica*—eventually marking up 439 species. With such study completed, he was more properly ready to engage in Stillingfleet's proposal and in the following year, 1766, he kept a detailed record of his observations in a volume to which he gave the name 'Flora Selborniensis, with some coincidences of the coming, & departure of birds of passage, & insects, & the appearing of Reptiles'. Inspection of this volume confirms White's bold hope, for recorded alongside the records of plants and creatures are data about the weather.

More significantly, however, for the future appearance of *Selborne* was the stimulus this experience gave to his fortunes the following year. In late spring 1767 White was in London for almost two months. During this visit he met Thomas Pennant, most probably at the bookshop of his brother (Benjamin White), which was a meeting-place for enthusiasts of natural history. Already an established naturalist and author, Pennant had corresponded with Linnaeus since 1755, toured the continent, and was the author of *British Zoology* (1761–6), the standard text for the period. The consequences of this meeting for White were decisive: previously, his discussions about the natural scene had been restricted to local farmers and villagers (who often furnished him with information and specimens), to members of his family (notably his brother Thomas, with whom he botanized for bog plants in August 1767 and who in the 1780s wrote a series of articles on trees for the *Gentleman's Magazine*), and to the occasional academic colleague at Oxford. Now, not only had he met an author with a national reputation, but also, through Pennant's good offices, was introduced to a young man on the edge of an international reputation—Joseph Banks, whose circumnavigation of the world (1768–72) and return with a vast magazine of trophies, principally specimens of flora and fauna never previously seen in London, was to prompt much applause and amazement. From this distance of time (and from our contemporary introductions to 'the great and the good'—at least through the medium of television), such encounters seem unimportant, even trivial; but that is to misunderstand White. Whereas in the previous decade he had discovered that his qualifications, academically, socially, financially, were insufficient to secure election as head of an Oxford college, now he discovered that his knowledge of natural history, which he somewhat discounted—much of his knowledge being common in the parish—was of value to the grand world. In particular Pennant, who was planning a new edition of *British Zoology*, quickly perceived that White could furnish reliable information concerning ornithology, and invited him to communicate with him. The invitation was accepted and White's first letter (of 10 August 1767) became later, in an edited form, Letter 10 of *Selborne*, with the date 4 August 1767.

From this simple beginning a more challenging consequence was to follow. The Linnaean programme and Stillingfleet's ideas had brought new thinking to some of the general questions that beset the relationship between natural history and husbandry. Alongside publication of Hudson's *Flora*, Pennant's own volumes, and many books about gardening and agricultural practice, there emerged in 1767 a work that is best described as a naturalist's ledger. With the title *The Naturalist's Journal*, it had been designed by Daines Barrington (see 'Biographical Notes' below) as a format to collect daily local records of the weather alongside observations of plants and creatures, so that 'from many such journals kept in different parts of the kingdom, perhaps the very best and accurate materials for a General Natural History of Great Britain may in time be expected, as well as many profitable improvements and discoveries in agriculture' (Preface to *The Naturalist's Journal*).

Late in 1767, at Pennant's suggestion, Barrington sent to Selborne a copy of this journal. Comparison of Barrington's proposal with Stillingfleet's hopes concerning the collection of coincidences shows that the architect of the journal (Barrington) was much more sceptical than Stillingfleet, and held more rigidly to the straightforward amassing of data. Perhaps Barrington's legal training was at work in this.

Whatever the case, the format appealed to White and he set to work with alacrity. Each page was ruled to offer eleven different kinds of entry daily, and on Friday, 1 January 1768, White began a process that was to occupy him for twenty-five

years—until the final entries of 15 June 1793, a mere ten
days before his death. The repetitiveness of the process that
required—regardless of whether there was anything really
interesting such as the arrival on 13 April 1768 of the house
swallow which was entered with obvious delight as 'Hirundo
domestica!!!'—the routine recording of temperature, barome-
tric pressure, wind direction, and weather conditions pleased
White's temperament, and within a few months he was in
London to talk with Barrington. The occasion was as significant as the meeting with Pennant the previous year, White
recording his delight in Barrington's 'candor, & affability'
(letter to Pennant of 16 June 1768). The following year, again
in London, White renewed his acquaintance and agreed to
initiate correspondence—which he honoured with a letter to
Barrington of 6 July 1769 (after editing this became *Selborne*
DB 1[2] of 30 June 1769). Within a few months Barrington fully
recognized White's abilities and proposed he prepare 'an
account of the animals in [his] neighbourhood' (DB 5, 12
April 1770). At first White made a hesitant response: he was
fully aware of the errors such a work might contain (see, for
example, TP 39 and TP 40 for the additional information and
corrections he was to supply to someone who *claimed* authority), and of the contrast between the limited opportunities for
observation and the paucity of information that might
accrue—especially when the observer was committed to accuracy (see DB 5).

At one level such hesitation suggests a false modesty (White
knew enough for any field naturalist, then or now, to learn
something from him), but at another it points to a distinctive
mark of his contribution—a commitment to produce not a
volume of names, descriptions, and synonyms, important
though he knew Linnaean systematic listings to be, but a
volume that focused on the 'life & manners of animals [which]
are the best part of Nat: history'. This approach, towards an
interest in *behaviour*, distinguishes White's contribution to the

---

[2] In the Introduction and Explanatory Notes to the text, the letters to
Thomas Pennant are designated TP and those to Daines Barrington DB, the
numbers referring to the published version.

study of natural history and precisely at the period of Barrington's invitation, it received endorsement from an unusual location—Gibraltar.

John White (a brother) was chaplain to the garrison at Gibraltar for over fifteen years. During the last three years of his tenure (1769–72), he pursued, under White's direction, a formal study of the local natural history. The tutelage was by correspondence (see P. G. M. Foster, 'The Gibraltar Correspondence of Gilbert White', *Notes and Queries*, 32 (1985), 227–36, 315–28, 489–500), and in return for books and advice White received cargoes of specimens. The identification of these specimens, birds, plants, insects, fish, set White the most daunting task of his career. Excited as he was, however, both by the accomplishments of his brother and by the flow of exotic specimens (the identification of which necessitated assistance from many specialist authorities in London), White found this kind of natural history, mere naming, of less appeal than his own enthusiasm for 'life & manners'. Admittedly, on John's return to England the two brothers were to spend a whole winter together at Selborne (1772–3) to review the many items John had sent from Gibraltar, a period in his life that later White was to claim as 'one of the most pleasant' (White to John, January 1774); and he was to continue to cajole and encourage his brother towards a 'Fauna Calpensis', a natural history of Gibraltar. All these varied ventures and meetings were both flattering and instructive; they included visits to Selborne in 1769 and 1770 by William Sheffield, Provost of Worcester College, who also proposed that John White prepare a *Fauna Gibraltarica*—see P. G. M. Foster, 'William Sheffield: Four Letters to Gilbert White', *Archives of Natural History*, 12 (1985), 1–21. It was the prescience of Barrington, however, that should be acclaimed as the most compelling factor in the eventual appearance of *Selborne*. As indicated above, White's hesitation in providing an immediate answer to Barrington's had first to be stilled; and again, perhaps Barrington's legal training was of value (in, for example, coaxing responses from witnesses). Initially he tolerated White's indecision, and then made a bold suggestion, that the Selborne naturalist prepare monographs on the behaviour of his most favoured creatures—swallows, martins, and swifts. The monographs would be in the form of letters which he (Barrington) would read to a meeting of his fellow members of the Royal Society—a customary mode of the period for sharing new learning. This proposal was much to White's liking, and the outcome was decisive. Barrington presented the first monograph on 10 February 1774; the resulting acclaim convinced White he had something to offer the world, and within a few weeks he was writing to his brother John (now vicar at Blackburn, Lancashire): 'As to my letters [by which he meant his correspondence with Pennant and Barrington], they lie in my cupboard very snug: if you will correct them ... I will publish' (White to John White, 29 March 1774).

The journey that had begun with the fall of gossamer in 1741, in an attempt to meet the mystery of Providence in the natural world, was to become public and be shared with a wider audience. Some rethinking was to follow, notably a decision to add to the volume a history of the antiquities of the parish, and there were to be the inevitable delays (primarily family matters associated with living and dying); and later, as the whole volume began to take shape, the incorporation of additional letters made up from material in his 'Naturalist's Journal'. These were White's entries on the printed pro-forma diary sheets devised in 1767 by Daines Barrington, which had ruled columns for naturalists to record their observations in eleven categories. The combination of sources reflects White's personality. He had no wish to constrain his observations in an inert tome of classification: what he wished to write had to breathe, to maintain a responsiveness to his own continuing delight in the rich varieties of the English rural and pastoral scene—even if publication was to be delayed. It is to that delight, which has sustained countless readers since 1789, to which we now turn.

Selborne *itself, and What it Offers*

Selborne made a new and highly original contribution to the writing of natural history. At its publication on 1 November 1788 (convention permitting a book issued so late in the year to carry the date of publication as that of the year following), copies were sent to relatives and friends—including John Mulso at Winchester, who arranged for it to be included in the cathedral library. A number of reviews quickly followed, Thomas White in the *Gentleman's Magazine* (1789), 60–3, 144–6, being especially discreet in his assessment of his brother's work. It was the reviewer in the *Topographer*, 1: 1 (1789), however, who caught the manner and tone of White's volume with especial insight and grace:

The book is not a compilation from former publications, but the result of many years' attentive observations to nature itself, which are told not only with the precision of a philosopher, but with that happy selection of circumstances, which mark the *poet*. Throughout therefore not only the understanding is informed, but the imagination is touched.

Several phrases of this judgement point to the distinctiveness of White's achievement. From the perspective of *Selborne*, it can be seen that earlier books about natural history fall into one of three broad groupings—travel books, guidebooks, or compilations—each of which contributes something to the character of White's work. Consider the last: compilations are best understood as encyclopaedias or dictionaries, arranged either by a systematic classification or by a paragraph or two of description. In a systematic classification an entry for, say, the sand-martin, *Riparia riparia*, might read:

Hirundo riparia: H. cinerea, gula albis .... Habitat in *Europae* collibus arenosis abruptis, foramine serpentino [Bank swallow, ash-coloured, with white throat and belly .... Found in Europe near steep sandy hills, with a serpentine nest-hole]. (Linnaeus, *Systema Naturae*, 10th edn., 1758)

In a descriptive passage this might occur:

The Sand Martin .... this is the le[a]st of the genus that frequents *Great-Britain*. The head and whole upper part of the body are mouse colored: the throat white, encircled by a mouse colored ring: the belly white: the feet smooth and black.
   It builds in holes in sand pits, and in the banks of rivers. It makes its nest of hay, straw, &c. and lines it with feathers: it lays five or six eggs, which are white, as are all those of this tribe [by which the

author means the swallow family]. (Thomas Pennant, *British Zoology*, vol. ii, 1768)

Comparison of either of these approaches with what White writes about the sand-martin (DB 20) is instructive, for it is immediately apparent that White possessed a quite different approach to natural history. At one level this difference concerns the function of writing, but at another it reveals a fundamental difference over natural history as a form of knowledge. Although both Linnaeus and Pennant were active observers of the natural scene, they aimed in their principal publications to subjugate their knowledge to a particular mode, whether systematic (Linnaeus) or descriptive (Pennant). Further, because their field was almost the entire world (Pennant, for example, as well as *British Zoology*, published *Arctic Zoology* (1784–7) and *Indian Zoology* (2nd edn., 1790) although he visited neither region), both men were forced to rely on informants whose standards of accuracy were not always reliable. White's approach was diametrically different. Recognizing the value of a systematic approach once sufficiently accurate data were available, he averred the importance of local studies, for it was those who studied 'only one district' who were 'much more likely to advance natural knowledge than those that grasp at more than they can possibly be acquainted with' (DB 7; see also DB 10 and DB 40). Restrictive as such a view may appear, it was born of a familiarity with much more than the parish of Selborne. White possessed personal knowledge of the whole of central southern England, and of habitats north and east (to Rutland, the Isle of Ely, and London) and west (across Salisbury Plain and beyond—into Devon). Also, he regularly surveyed the exhibitions and displays in London, and shared acquaintance with travellers across wider horizons: Pennant, who toured Wales and Scotland (twice), as well as the continent; James Gibson, who saw military action at Quebec (1759); John White, who sent cargo after cargo of specimens from Gibraltar; and Joseph Banks, a circumnavigator of the entire globe. In addition, there is compelling evidence, much of it present in *Selborne*, that he read widely not only in classical and modern (English) literature, but also in the principal authors of the age concerned with natural history—Adanson, Aikin, Brisson, Buffon, Derham, Edwards, Ellis, Fothergill, Geoffroy, Hales, Hasselquist, Huxham, Kalm, Lightfoot, Miller, Réaumur, Stillingfleet, Swammerdam. All these experiences, far from distracting White from the life of his own parish, served to emphasize the strength and accumulated authority of his own painstakingly incremental and personal knowledge of the natural history of his own domain.

Faced, however, with a growing confidence in his capacity to publish, a decision had to be made about the best form of communication. If 'bare descriptions, and a few synonyms' (DB 10) were to be eschewed, the real letters written to Pennant and Barrington (when edited), even with the addition of a specially written context (TP 1–9), would scarcely provide sufficient material. Nevertheless, the reworking of letters to other correspondents (for example, DB 30 is taken directly from a letter to John White of 12 August 1775) and the writing-up in letter-form of suitable entries from his 'Naturalist's Journal' (see DB 56, note to 'distance') might flesh out the core correspondence to create a modest volume, especially if, as a second part, information about the antiquities of the parish was included.[3] The decision to include antiquities was not entirely a matter of extending the available material: it also preserved one kind of influence. Previous volumes in the century in the category of natural history guidebooks had taken a model from Robert Plot, whose *Natural History of Oxfordshire, Being an Essay toward the Natural History of England* (1677) had been followed by, for example, *The Natural History of Lancashire, Cheshire, and the Peak, in Derbyshire: with an Account of the Antiquities in those Parts* (1700) by Charles Leigh, and by

---

[3] The relocation of material from one correspondent to another and the reworking of entries in the 'Naturalist's Journal' as letters may lead the present-day reader to question White's integrity; but the conventions of communication differ from age to age and White's process in preparing *Selborne* (an outcome of literary creation) is best seen as analogous to the production of a modern wildlife film—hours of filming lead, after editing, to no more than a few minutes' transmission, and yet the integrity of the final result is accepted: in such terms also do we accept the truth of *Selborne*.

A Natural and Historical Account of the Islands of Scilly (1750) by
Robert Heath. Such works, to take Heath as a guide, endeavoured to offer a comprehensive description of the chosen
territory, of its soil, culture, situation, produce, rarities, trade,
inhabitants, government, customs, antiquities, and so on.
Comparison of this kind of inclusivity with a complete *Selborne*
(the natural history portion *and* the antiquities) shows that
White covers the same territory, but with one crucial difference: that of communicative sequencing and tone. Natural
history in guidebook mode concentrated on the report of
information organized in a coherent, rational, and logical
fashion. Plot, for instance, devoted chapters, in sequence, to:
'Of the Heavens and Air'; 'Of the Waters'; 'Of the Earths';
'Of Stones'; 'Of Formed Stones [fossils]'; 'Of Plants'; 'Of
Brutes [Creatures]'; 'Of Men and Women'; 'Of Arts [things
made by man: roads, carts, instruments . . .]'; 'Of Antiquities'.
An ordering of this kind is far removed from what *Selborne*
offers. White was committed to 'life and manners', to capturing
the quotidian intricacies of the natural world.

Further, and concomitant with this predilection for 'behaviour', is a sure grasp of immediacy. It is a quality of writing
that informs several aspects of *Selborne* and that, in terms of
earlier models, is carried over from natural history as travel
books. Volumes in the genre known to White begin with John
Ray's *Observations . . . made on a Journey through Part of the Low
Countries, Germany, Italy, and France* (1673), and include M.
Adanson, *A Voyage to Senegal* (1759), Frederick Hasselquist,
*Voyages and Travels in the Levant . . . containing Observations in
Natural History* (1766), Peter Kalm, *Travels into North-America;
containing its Natural History* (1770), and Peter Osbeck, whose
Voyage to China and the East-Indies (1771) was revised by White
for its English style. The distant places these works described,
together with the restricted information such volumes contained, emphasized to White the value of his own focus on a
single, if restricted, locale; and yet, at the same time, they also
confirmed a commitment to what is best termed *communicative
presence*. A travel writer customarily writes about the new, the
unusual, the strange. Such a focus contributes powerfully to
the liveliness of travel writing, and it is a mark of White's

originality that *Selborne* communicates a similar quality—not by describing foreign, mysterious, and uncharted regions, but by a concentration on the homely and (literally) the parochial. This quality of White's volume flickers brightly across the whole of *Selborne*, and depends—in ways that are frequently overlooked—on the decision to retain in the published volume the letter form of White's first communications with Pennant and Barrington. Letters are communicative acts in a way that guidebooks or dictionaries are not. They are written for a specific audience, whom the writer knows wishes to share what is to be communicated. In consequence, the writer declines to tell us, as in part dictionaries and guidebooks do, what we already know; on the contrary, as the correspondence develops, there is established both a shared knowledge and a willingness to take risks—that is, to admit limitations (see TP 21), to acknowledge errors (see TP 40; DB 8), to provide additional information to that already furnished (see TP 29, note to 'me'; DB 52). In all these ways, and in others (for example, White's homely comparisons—TP 35; recognition of the value of agricultural improvements—DB 40; admission of childish credulity—DB 45), White reveals the intensely human dimension of his personality and the artless accomplishment that is *Selborne*. Indeed, it is this very artlessness with which the modern reader can most readily identify. In terms of both the actual observations White writes about and the manner in which they are communicated, *Selborne* acknowledges life to be a richly patterned, variable kaleidoscope. Despite the horrors of the time— and for White these embraced the loss of the American colonies, war with the French, the revolution that threatened across the Channel and, at home, the devastation of enclosures, the beginnings of the industrial revolution, poor harvests and dreadful winters, riots in London (the Gordon Riots of 1780)— investigations into the natural world and the communication of those investigations (especially where they prompted indeterminate speculation) were inexhaustible. In this context White's approach to knowledge about natural history must be classed as supremely optimistic. Uncertain and threatening as the world seemed, he believed absolutely that man and his position in it could be sustained, if not explained fully.

Central to this belief, to put it boldly, was his absorption in the rich variety of creaturely behaviour, including human. It is this that places at the centre of *Selborne* the 'State of the Parish' (TP 5, White's note 7). Study of bees and birds rested ultimately on the birds and the bees: in face of human and natural disaster, Selborne as a community was holding its own, and even growing. In this sense White is profoundly modern, expressing a notion of knowledge that is evolutionary. He knew himself as a creature to be fallible, enriched by what he could comprehend, humbled before Providence by what he failed to understand. As he moved through the beeches on the hanger, or glimpsed the first annual swoop of a hirundine, he knew—as Keats was to put it a generation later—that 'the poetry of earth is never dead' and that the birth and rebirth of the natural scene was secure in the distinctively English countryside, temperate, sweet.

NOTE ON THE TEXT

THE text of *Selborne* (1789), the sole edition published in White's lifetime, was printed in London by T. Bensley of Bolt Court, Fleet Street. White himself scrutinized the proofs and prepared an index, a task he found little to his liking, seeing it as 'an occupation full as entertaining as that of darning of stockings, tho' by no means so advantageous to society' (*Selborne*, ed. T. Bell (London: Van Voorst, 1877), ii. 168). This lack of enthusiasm for the task, together with his distance from London, meant that responsibility for overseeing the detail of publication fell to members of White's family resident in the capital—with, sadly, some unfortunate results: for example, the natural history portion of the text has fifteen printer's errors, some letters are misnumbered (as is one page), an additional section to DB 41 is misplaced, and the 'More Particulars . . . Tortoise' (see DB 50) is printed after the 'Antiquities'.

Such misfortunes have led many editors to take a text from the third (English) edition of 1813—the first to include editorial notes (by the Revd John Mitford of Benhall, Suffolk). In the present edition, however, priority has been given to the 1789 text. This decision has rested primarily on the availability at Selborne of White's copy text. Consultation of this manuscript material has revealed much about White's procedures during the preparation of the volume, and the fruits of this consultation occur both in the introduction to the present edition and in notes—see, for example, note to p. 118 (DB 8), '*SIR*'.

Yet a text of 1789, even when corrected by the careful scrutiny of White's manuscript intentions, would seem strange to the eyes of a modern reader, and certain modifications have therefore been effected. These modifications, substantial as they may seem when listed, are solely matters of presentation and printing convention; they detract nothing from White's literary style or scientific acumen. They include: the conversion of letter numbering from roman numerals to arabic; the simplification of the printer's use of inverted commas—at the beginning of every line, for example, in long prose or poetry quotations; regularizing the use of the apostrophe; the expansion of the 'æ' and 'œ' diphthongs to give, for example, 'grallae' for 'grallæ' and 'foetidus' for 'fœtidus' (although I give 'œconomists' as 'economists'); the printing of all proper nouns in roman and not, as in eighteenth-century fashion, in italic type; the avoidance of distracting period spellings—for example, 'embrio' has become 'embryo', 'kestril' has become 'kestrel', although in the interests of historical awareness some differences from modern conventions are retained (for example, White uses both 'chimnies' and 'chimneys', 'wigeon' and 'widgeon', and I retain spellings such as 'scissars' for 'scissors' and 'chuse' for 'choose'). In White's many quotations from classical and English literature line numbers are omitted. This last has been an instructive study: in many instances, because of familiarity to an eighteenth-century readership, White provides no citations of author or of text; in a few instances he gives full details, including line numbers. For a modern reader, however, many of White's epigraphs and other quotations are unfamiliar and I provide details of author and title; line numbers are omitted entirely since those given by White differ markedly from numbering used in modern editions, and the provision of book or volume numbers has seemed a sufficient guide for readers wishing to pursue White's contextual selections.

Finally—plates, poems, and index: the 1789 edition contained a frontispiece showing a panoramic view of Selborne with the hanger behind, several plates, two poems ('The Naturalist's Summer-Evening Walk' of TP 24, and White's untitled verses on the autumn crocus at the conclusion of DB 41), and White's index. This last is rudimentary and I now provide an enlarged index which gives modern forms for many place names, which lists all plant and animal species, and which offers comprehensive points of entry to the many authors White discusses. As for plates, it has seemed inappropriate to include them within the present format: they were designed for a quarto edition and the necessary reduction now required would be a disservice to White's artist, Samuel H.

Grimm, and to his engravers, D. Lerpinière and P. Mazell. In compensation I include, first, a number of illustrations by Edmund H. New from the 1901 Grant Allen edition, which were used in the original World's Classics edition of 1902; and second, 'The Invitation to Selborne'. These engaging lines, the most extended of White's several poems, were first drafted and sent by White to his lifelong friend, John Mulso, in 1751 as a real invitation to visit Selborne: it is an invitation we can still accept today.

SELECT BIBLIOGRAPHY

EDITIONS OF *SELBORNE* (1789)

The popularity of *Selborne* has been such that, although at the time of first publication (1789) it was available in one volume under the title 'The Natural History and Antiquities of Selborne', most subsequent editions have entirely omitted the 'Antiquities'. Many of these editions, especially those of last century, not only included extracts from White's other writings, but also developed distinctive modes of presentation: for example, some editors offered only selections from *Selborne*; others divided up the letters into topics, grouped the passages on one topic together, and then re-presented the whole as a one-volume encyclopaedia; and others interleaved the letters addressed to Pennant with those addressed to Barrington so as to provide a continuous chronology (determined solely by the putative dates of the letters). Instructive as these editions are, particularly as examples of editorial intervention, they thwart White's artistic (literary) purposes and present readers with a volume far-removed from White's intentions. All these editions, and many more, are described in Edward A. Martin, *A Bibliography of Gilbert White* (Folkestone: Dawsons, 1897; rev. edn. 1934, reprinted with additions 1970). Some of the most notable editions, which now number over 200 and rank *Selborne* not just as the most published scientific text, but also as a literary classic rivalled only by works such as the Bible, Shakespeare's plays, and Bunyan's *Pilgrim's Progress*, are given below—by editor and date of publication:

Thomas Bell (London: Van Voorst, 1877)—a comprehensive edition that includes some correspondence with family and friends.

Grant Allen (London: Bodley Head, 1899)—gives in an appendix MS marginalia by Samuel Taylor Coleridge taken from an 1802 edn. now in the British Library.

Bowdler Sharp (London: S. T. Freemantle, 1900)—incorporates first publication in its entirety of the 'Garden Kalendar'.

E. M. Nicholson (London: Thornton Butterworth, 1929)—with an excellent introduction.

James Fisher (London: Cresset, 1947)—notes confined mainly to matters of interest to the ornithologist.

Editions have also been published in German (1792), Danish (1951), Swedish (1963), Japanese (1949; 1958). There have been many illustrated editions, of which the following are distinctive:

John Burroughs (London: Macmillan, 1895)—exquisite period illustrations by Clifton Johnson.

Richard Kearton (London: Cassell, 1902)—fine photographic
illustrations.

H. J. Massingham (London: Nonesuch, 1938)—illustrated by Eric
Ravilious.

John Lewis (London: Lutterworth, 1951)—drawings by John Nash.

W. S. Scott (London: Folio Society, 1962)—drawings by John Piper.

June E. Chatfield (Exeter: Webb & Bower, 1981)—coloured illustrations taken from contemporary eighteenth-century works.

As for the 'Antiquities', it was not separately published until the
middle of this century—by Falcon Press (London, 1950), with a fine
set of notes by W. Sidney Scott. A more recent edition, with an
introduction by June E. Chatfield, was published by Gresham Books
(Henley-on-Thames, 1982).

OTHER WRITINGS

Although extracts from some of White's journals will be found in early
editions of *Selborne*, the complete text of the 'Garden Kalendar'
(1751–71) and the 'Calendar of Flora' (1766), together with the
annual volumes of his 'Naturalist's Journal' (1768–93), is best
consulted in *The Journals of Gilbert White 1751–1793*, ed. Francesca
Greenoak, 3 vols. (London: Century Hutchinson, 1986–9). Two
facsimile editions of works by White show his elegant and marvellously clear handwriting: *Garden Kalendar*, ed. John Clegg (London:
Scolar Press, 1975), and *A Nature Calendar*, an edition of the 'Calendar
of Flora', ed. William Mark Webb (London: Selborne Society, 1911).

For a delightful collection of White's observations on his tortoise
Timothy, see Sylvia Townsend Warner, *The Portrait of a Tortoise*
(London: Chatto & Windus, 1946).

BIOGRAPHICAL AND CRITICAL WORKS

Biographical

White's biography begins with remarks made by John White (a
nephew—son of White's brother Benjamin) that were first printed in
the 1802 edition of *Selborne*. Progressively during the following years
more material became available, and several editions announced the
inclusion of extracts from White's journals and, also, the publication

for the first time of letters from White to one or other of his correspondents. This additional material, especially the correspondence with family and friends, became so great that in the mid-1870s Thomas Bell, who lived at Wakes for thirty-eight years, was able to assemble it into an entire second volume to accompany his edition, *The Natural History and Antiquities of Selborne*, 2 vols. (London: Van Voorst, 1877). Following this, in the first decade of the present century, Rashleigh Holt-White, a great-grand nephew of White's, prepared two important texts: first, *The Life and Letters of Gilbert White of Selborne*, 2 vols. (London: John Murray, 1901); and second, *The Letters to Gilbert White from his Intimate Friend and Contemporary, The Rev. John Mulso* (London: Porter, 1907).

Numerous later volumes have appeared. The most helpful are: Walter S. Scott, *White of Selborne and his Times* (London: Westhouse, 1946), which concentrates on academic and social matters unrelated to natural history; R. M. Lockley, *Gilbert White* (London: Witherby, 1954), written with the eye of a naturalist; Cecil S. Emden, *Gilbert White in his Village* (London: Oxford Univ. Press, 1956); and Anthony Rye, *Gilbert White & his Selborne* (London: Kimber, 1970), a celebration for the 250th anniversary of White's birth. For a more recent, popular biography, see Richard Mabey, *Gilbert White* (London: Century Hutchinson, 1986), and for a work that emphasizes White's development as a natural historian, see Paul G. M. Foster, *Gilbert White and his Records: A Scientific Biography* (London: Christopher Helm, 1988).

Critical

The comprehensive range of *Selborne*—the way it takes account of the whole of the natural history of the parish, and much else besides—has created difficulties for modern critics and the standard text remains Walter Johnson, *Gilbert White: Pioneer, Poet, and Stylist* (London: John Murray, 1928). More recent criticism has concentrated on aspects of White's work and is, therefore, available mainly as chapters of books devoted to a theme that embraces White's interests, and in periodicals. In the former category there are excellent chapters in W. J. Keith, *The Rural Tradition* (Hassocks, Sussex: Harvester, 1975), and in Paula R. Backscheider (ed.), *Probability, Time, and Space in Eighteenth-Century Literature* (New York: AMS Press, 1979). As for periodical literature, Jack Stillinger wrote a fine article, 'Gilbert White to Thomas Pennant: Two Original Letters at Harvard', *Harvard Library Bulletin*, (1957), vol. 11 (1957), 303–16. The following are also useful: Charles F. Mullet, '*Multum in parvo*: Gilbert White of Selborne', *Journal of the History of Biology*, 2 (1969), 363–89; G. I. Meirion-Jones, 'The Wakes, Selborne: An Architectural Study', *Proceedings of the Hampshire Field Club and Archaeological Society*, 39 (1983), 145–69; Arthur Sherbo, 'The English Weather, *The Gentleman's Magazine*, and the Brothers White', *Archives of Natural History*, 12 (1985), 23–9; J. E. Chatfield, 'Likenesses of the Reverend Gilbert White', *Proceedings of the Hampshire Field Club and Archaeological Society*, 43 (1987), 207–17. Foster, *Gilbert White and his Records*, cited above, gives bibliographic details for articles on White's correspondence with his brother at Gibraltar, on letters from William Sheffield that led White to understand better the principles of the Linnaean system, on White's botanical records, on Barrington's annotation of White's journals, and on White's informant concerning moose in North America; and Paul G. M. Foster, 'Approaches to the Study of Gilbert White (1720–1793): Or, a Balloon over Selborne', *Archives of Natural History*, 17 (1990), 299–314, offers an overview of current attitudes to White scholarship.

For an appealing monograph that cites White's observations on a creature he studied all his life, see Margaret Grainger and Richard Williamson, *The Nightjar: Yesterday and Today* (Chichester, Sussex: Bishop Otter Trustees, 1988).

GENERAL WORKS

These fall into two groups—contemporary with White, and contemporary with White's present readership. In the former group the most important are: John Ray, *The Wisdom of God Manifested in the Works of Creation* (1691); W. Derham, *Physico-Theology: or, a Demonstration of the Being and Attributes of God, from his Works of Creation* (1713); Benjamin Stillingfleet, *Miscellaneous Tracts Relating to Natural History* (1759); and Daines Barrington, *Miscellanies* (1781). Thomas Pennant's several works (see Biographical Notes) should also be consulted. For detail about White's meteorological mentor, see John Kington (ed.), *The Weather Journals of a Rutland Squire: Thomas Barker of Lyndon Hall* (Oakham: Rutland Record Society, 1988), and for vital works illuminating White's poetic inheritance see Virgil, *Georgics*, and James Thomson, *The Seasons* (1730)—both of which are readily available in modern editions.

In the latter group, works of our own time, the following provide excellent contextual frames to which White can be related:

David Allen, *The Naturalist in Britain* (London: Allen Lane, 1976).
Linda Colley, *Britons: Forging the Nation 1707–1837* (New Haven, Conn.: Yale Univ. Press, 1992).

John Prest, *The Garden of Eden: The Botanic Garden and the Re-creation of Paradise* (New Haven, Conn.: Yale Univ. Press, 1981).

James Sambrook, *The Eighteenth Century: The Intellectual and Cultural Context of English Literature 1700–1789* (London: Longman, 1986).

Keith Thomas, *Man and the Natural World: Changing Attitudes in England 1500–1800* (London: Allen Lane, 1983).

Lastly, a monograph published to accompany 'Cultural Landscapes: An Exhibition on the Bicentenary of Gilbert White's *Natural History of Selborne*' (Stanford, Calif.: Stanford Univ. Libraries, 1989), with essays by W. B. Carnochan and Elizabeth Heckendorn Cook, is contemporary in the best of both senses and values White's world both for what it was and for what it still is, today.

A CHRONOLOGY OF
GILBERT WHITE

1720   18 July: Born at Selborne, Hampshire, eldest child of John
        White (1688–1758).

1728   Death of grandfather, the Reverend Gilbert White (vicar of
        Selborne from 1681); White's parents, after living near
        Guildford and in East Harting, settle at Wakes, Selborne.

1730   Plants oak and ash at Wakes.

1736   Makes first extant natural history observations while on
        holiday in Rutland.

1739   School at Basingstoke; death of mother (17 December).

1740   Begins studies at Oriel College, Oxford.

1741   21 September: Observes fall of gossamer.

1742   Spends summer months with relatives in Rutland.

1743   Awarded BA degree—presented with copy of Pope's *Iliad*.

1744   Elected Fellow of Oriel.

1746   Attends, for six months, to estate of relative in Isle of Ely.

1747   27 April: Made deacon in Christ Church, Oxford; first
        curacy at Swarraton, Hampshire; attack of smallpox.

1749   11 March: Ordained priest in Spring Gardens, London.

1750   Tour to Devon—returns with sea-kale.

1751   Begins 'Garden Kalendar' and maintains detailed records
        until 1767; curate (briefly) at Selborne[1]

1752   Dean of Oriel; Junior Proctor in the University.

1753   Visits Bristol Hot-well (and again in 1755); curate at Durley,
        Hampshire.

1755   Curate at West Dean, Wiltshire.

1757   Candidate for Provostship at Oriel.

1758   29 September: Death of father.

1760   Tour to Rutland—White's most extended absence from
        Selborne.

[1] White was curate at Selborne on three further occasions—in 1756-7,
1758-9, and from 1784 to his death.

1761   Completes ha-ha at Wakes and builds fruit-wall;
        curate at Farringdon, Hampshire.

1763   Inherits Wakes.

1766   Keeps 'Flora Selborniensis'.

1767   Meets Joseph Banks in London—to pursue 'Ornithological
        Converse'; first letter to Thomas Pennant; reports discovery
        of harvest mouse.

1768   Begins 'Naturalist's Journal' and enters a record of
        observations (almost daily) until 15 June 1793.

1769   First letter to Daines Barrington; Oxford naturalists,
        William Sheffield and Richard Skinner, visit Selborne;
        receives first specimens from brother (John), chaplain at
        Gibraltar, and advises him about writing natural history.

1772   John (brother) returns from Gibraltar and spends winter
        1772–3 at Selborne discussing natural history.

1774   10 February: First paper read at meeting of Royal Society.

1775   16 March: Second paper read at Royal Society.

1776   July: Samuel H. Grimm at Selborne—takes twelve views in
        twenty-four days for the projected *Selborne*.

1777   Thomas (brother) elected FRS; begins work on new parlour
        at Wakes—completed 1780.

1784   16 October: Observes Blanchard's balloon pass over
        Selborne.

1788   1 November: *Selborne* published.[2]

1791   21 September: Death of Reverend John Mulso—active
        correspondent since undergraduate days.

1793   26 June: Dies at Wakes; buried at Selborne in the
        churchyard on north side of chancel (at his own wish).

[2] By convention books published in the final months of one year were
permitted to carry a publication date of the following year—hence, *Selborne*
1789.

The Invitation to Selborne

See Selborne spreads her boldest beauties round,
The varied valley, and the mountain ground,
Wildly majestic! what is all the pride
Of flats, with loads of ornament supply'd;
Unpleasing, tasteless, impotent expense,
Compared with nature's rude magnificence.

   Arise, my stranger, to these wild scenes haste;
The unfinish'd farm awaits your forming taste;
Plan the pavilion, airy, light, and true;
Thro' the high arch call in the length'ning view;
Expand the forest sloping up the hill;
Swell to a lake the scant, penurious rill;
Extend the vista, raise the castle mound
In antique taste with turrets ivy-crown'd;
O'er the gay lawn the flow'ry shrub dispread,
Or with the blending garden mix the mead;
Bid China's pale, fantastic fence delight;
Or with the mimic statue trap the sight.

   Oft on some evening, sunny, soft, and still,
The Muse shall lead thee to the beech-grown hill,
To spend in tea the cool, refreshing hour,
Where nods in air the pensile, nest-like bower;*
Or where the hermit hangs the straw-clad cell,†
Emerging gently from the leafy dell;
By Fancy plann'd; as once th' inventive maid
Met the hoar sage amid the secret shade;
Romantic spot! from whence in prospect lies
Whate'er of landscape charms our feasting eyes;
The pointed spire, the hall, the pasture-plain,
The russet fallow, or the golden grain,

---

  * A kind of an arbour on the side of a hill.

  † A grotesque building, contrived by a young gentleman, who used on occasion to appear in the character of a hermit.

The breezy lake that sheds a gleaming light,
Till all the fading picture fail the sight.

Each to his task; all different ways retire;
Cull the dry stick; call forth the seeds of fire;
Deep fix the kettle's props, a forky row,
Or give with fanning hat the breeze to blow.
Whence is this taste, the furnish'd hall forgot,
To feast in gardens, or th' unhandy grot?
Or novelty with some new charms surprises,
Or from our very shifts some joy arises.
Hark, while below the village-bells ring round,
Echo, sweet nymph, returns the soften'd sound;
But if gusts rise, the rushing forests roar,
Like the tide tumbling on the pebbly shore.

Adown the vale, in lone, sequester'd nook,
Where skirting woods imbrown the dimpling brook,
The ruin'd Convent lies; here wont to dwell
The lazy canon midst his cloister'd cell;*
While papal darkness brooded o'er the land,
Ere Reformation made her glorious stand:
Still oft at eve belated shepherd-swains
See the cowl'd spectre skim the folded plains.

To the high Temple would my stranger go,†
The mountain-brow commands the woods below;
In Jewry first this order found a name,
When madding Croisades set the world in flame;
When western climes, urg'd on by Pope and priest,
Pour'd forth their millions o'er the deluged East:
Luxurious knights, ill suited to defy
To mortal fight Turcéstan chivalry.

Nor be the Parsonage by the muse forgot;
The partial bard admires his native spot;
Smit with its beauties, loved, as yet a child,
(Unconscious why) its scapes grotesque, and wild.

---

* The ruins of a priory, founded by Peter de Rupibus, Bishop of Winchester
 † The remains of a preceptory of the Knights Templars; at least it was a
farm dependent upon some preceptory of that order. I find it was a preceptory,
called the Preceptory of Sudington; now called Southington.

High on a mound th' exalted gardens stand,
Beneath, deep valleys scoop'd by Nature's hand.
A Cobham here, exulting in his art,
Might blend the General's with the Gardener's part;
Might fortify with all the martial trade
Of rampart, bastion, fosse, and palisade;
Might plant the mortar with wide threat'ning bore,
Or bid the mimic cannon seem to roar.

Now climb the steep, drop now your eye below,
Where round the blooming village orchards grow;
There, like a picture, lies my lowly seat,
A rural, shelter'd, unobserv'd retreat.

Me far above the rest Selbornian scenes,
The pendent forest, and the mountain-greens,
Strike with delight; there spreads the distant view,
That gradual fades till sunk in misty blue:
Here Nature hangs her slopy woods to sight,
Rills purl between, and dart a quivering light.

THE NATURAL HISTORY
OF SELBORNE

In a Series of Letters addressed to

THOMAS PENNANT, Esq.

and

THE HON. DAINES BARRINGTON

Τρηχεῖ', ἀλλ ἀγαθὴ κουροτρόφος. οὔτι ἔγωγε
Ἧς γαίης δύναμαι γλυκερώτερον ἄλλο ἰδέσθαι.

(Homer, *Odyssey*)

Tota denique nostra illa aspera, et montuosa, et fidelis,
et simplex, et fautrix suorum regio.

(Cicero, *Oratio pro Cn. Plancio*)*

ego Apis Matinæ
More modoque
Grata carpentis . . . per laborem
Plurimum . . .

(Horace)

Omnia benè describere, quæ in hoc mundo, a Deo facta,
aut Naturæ creatæ viribus elaborata fuerunt, opus est
non unius hominis, nec unius ævi. Hinc *Faunæ & Floræ
utilissimæ*; hinc *Monographi* præstantissimi.

(Scopoli)*

ADVERTISEMENT

THE Author of the following Letters takes the liberty, with all proper deference, of laying before the public his idea of *parochial history*, which, he thinks, ought to consist of natural productions and occurrences as well as antiquities. He is also of opinion that if stationary* men would pay some attention to the districts on which they reside, and would publish their thoughts respecting the objects that surround them, from such materials might be drawn the most complete county-histories, which are still wanting in several parts of this kingdom, and in particular in the county of Southampton.*

And here he seizes the first opportunity, though a late one, of returning his most grateful acknowledgments to the reverend the President and the reverend and worthy the Fellows of Magdalen College in the university of Oxford, for their liberal behaviour in permitting their archives to be searched by a member of their own society,* so far as the evidences therein contained might respect the parish and priory of Selborne. To that gentleman also, and his assistant, whose labours and attention could only be equalled by the very kind manner in which they were bestowed, many and great obligations are also due.

Of the authenticity of the documents above-mentioned there can be no doubt, since they consist of the identical deeds and records that were removed to the College from the Priory at the time of its dissolution; and, being carefully copied on the spot, may be depended on as genuine; and, never having been made public before, may gratify the curiosity of the antiquary, as well as establish the credit of the history.

If the writer should at all appear to have induced any of his readers to pay a more ready attention to the wonders of the Creation, too frequently overlooked as common occurrences; or if he should by any means, through his researches, have lent an helping hand towards the enlargement of the boundaries of historical and topographical knowledge; or if he should have thrown some small light upon ancient customs and

manners, and especially on those that were monastic; his
purpose will be fully answered. But if he should not have been
successful in any of these his intentions, yet there remains this
consolation behind—that these his pursuits, by keeping the
body and mind employed, have, under Providence, contrib-
uted to much health and cheerfulness of spirits, even to old
age: and, what still adds to his happiness, have led him to the
knowledge of a circle of gentlemen whose intelligent communications, as they have afforded him much pleasing
information, so, could he flatter himself with a continuation of
them, would they ever be deemed a matter of singular satisfaction and improvement.

GIL. WHITE.

Selborne,
January 1st, 1788.

LETTERS

TO

THOMAS PENNANT, Esq.

Gilbert White's house

Letter 1

THE parish of SELBORNE lies in the extreme eastern corner of the county of Hampshire, bordering on the county of Sussex, and not far from the county of Surrey; is about fifty miles south-west of London, in latitude 51, and near midway between the towns of Alton and Petersfield. Being very large and extensive it abuts on twelve parishes, two of which are in Sussex, viz. Trotton and Rogate. If you begin from the south and proceed westward the adjacent parishes are Emshot, Newton Valence, Faringdon, Harteley Mauduit, Great Ward le ham, Kingsley, Hedleigh, Bramshot, Trotton, Rogate, Lysse, and Greatham. The soils of this district are almost as various and diversified as the views and aspects. The high part to the south-west consists of a vast hill of chalk, rising three hundred feet above the village; and is divided into a sheep down, the high wood, and a long hanging* wood called The Hanger. The covert of this eminence is altogether beech, the most lovely of all forest trees, whether we consider its smooth rind or bark, its glossy foliage, or graceful pendulous boughs. The down, or sheep-walk, is a pleasing park-like spot, of about one mile by half that space, jutting out on the verge of the hill-country, where it begins to break down into the plains, and commanding a very engaging view, being an assemblage of hill, dale, wood-lands, heath, and water. The prospect is bounded to the south-east and east by the vast range of mountains called The Sussex Downs, by Guild-down near Guildford, and by the Downs round Dorking, and Ryegate in Surrey, to the north-east; which altogether, with the country beyond Alton and Farnham, form a noble and extensive outline.

At the foot of this hill, one stage or step from the uplands, lies the village, which consists of one single straggling street, three quarters of a mile in length, in a sheltered vale, and running parallel with The Hanger. The houses are divided from the hill by a vein of stiff clay (good wheat-land), yet stand on a rock of white stone, little in appearance removed from chalk; but seems so far from being calcarious, that it endures extreme heat. Yet that the freestone still preserves somewhat that is analogous to chalk, is plain from the beeches which descend as low as those rocks extend, and no farther, and thrive as well on them, where the ground is steep, as on the chalks.

The cart-way* of the village divides, in a remarkable manner, two very incongruous* soils. To the south-west is a rank clay, that requires the labour of years to render it mellow; while the gardens to the north-east, and small enclosures behind, consist of a warm, forward, crumbling mould, called black malm, which seems highly saturated with vegetable and animal manure; and these may perhaps have been the original site of the town;* while the woods and coverts might extend down to the opposite bank.

At each end of the village, which runs from south-east to north-west, arises a small rivulet: that at the north-west end frequently fails; but the other is a fine perennial spring, little influenced by drought or wet seasons, called Well-head.[1] This breaks out of some high grounds joining to Nore Hill, a noble chalk promontory, remarkable for sending forth two streams into two different seas. The one to the south becomes a branch of the Arun, running to Arundel, and so falling into the British channel; the other to the north, the Selborne stream, makes one branch of the Wey; and, meeting the Black-down stream at Hedleigh, and the Alton and Farnham stream at Tilford-bridge, swells into a considerable river, navigable at Godalming; from whence it passes to Guildford, and so into the Thames at Weybridge; and thus at the Nore into the German ocean.*

Our wells, at an average, run to about sixty-three feet, and when sunk to that depth seldom fail; but produce a fine limpid* water, soft to the taste, and much commended by those who drink the pure element, but which does not lather well with soap.

To the north-west, north and east of the village, is a range of fair enclosures, consisting of what is called a white malm, a sort of rotten or rubble stone, which, when turned up to the frost and rain, moulders to pieces, and becomes manure to itself.[2]

Still on to the north-east, and a step lower, is a kind of white land, neither chalk nor clay, neither fit for pasture nor for the plough, yet kindly for hops, which root deep into the freestone, and have their poles and wood for charcoal growing just at hand. This white soil produces the brightest hops.

As the parish still inclines down towards Wolmer-forest, at the juncture of the clays and sand the soil becomes a wet, sandy loam, remarkable for timber, and infamous for roads.* The oaks of Temple and Blackmoor stand high in the estimation of purveyors, and have furnished much naval timber; while the trees on the freestone grow large, but are what workmen call *shakey*, and so brittle as often to fall to pieces in sawing. Beyond the sandy loam the soil becomes an hungry lean sand, till it mingles with the forest; and will produce little without the assistance of lime and turnips.*

Letter 2

IN the court of Norton farm house, a manor farm to the north-west of the village, on the white malms, stood within these twenty years a broad-leaved elm, or wych hazel, *ulmus folio latissimo scabro* of Ray, which, though it had lost a considerable leading bough in the great storm in the year 1703,* equal to a moderate tree, yet, when felled, contained eight loads of timber; and, being too bulky for a carriage, was sawn off at seven feet above the butt, where it measured near eight feet in the diameter. This elm I mention to shew to what a bulk *planted elms* may attain; as this tree must certainly have been such from its situation.

In the centre of the village, and near the church, is a square piece of ground surrounded by houses, and vulgarly called The Plestor.* In the midst of this spot stood, in old times, a vast oak, with a short squat body, and huge horizontal arms extending almost to the extremity of the area. This venerable tree, surrounded with stone steps, and seats above them, was the delight of old and young, and a place of much resort in summer evenings; where the former sat in grave debate, while the latter frolicked and danced before them. Long might it have stood, had not the amazing tempest in 1703 overturned it at once, to the infinite regret of the inhabitants, and the vicar, who bestowed several pounds in setting it in its place again: but all his care could not avail; the tree sprouted for a time, then withered and died.* This oak I mention to shew to what a bulk *planted oaks* also may arrive: and planted this tree must certainly have been, as will appear from what will be said farther concerning this area, when we enter on the antiquities of Selborne.

On the Blackmoor estate there is a small wood called Losel's, of a few acres, that was lately furnished with a set of oaks of a peculiar growth and great value; they were tall and taper like firs, but standing near together had very small heads, only a little brush without any large limbs. About twenty years ago the bridge at the Toy, near Hampton Court,* being much decayed, some trees were wanted for the repairs, that were fifty feet long without bough, and would measure twelve inches diameter at the little end. Twenty such trees did a purveyor find in this little wood, with this advantage, that many of them answered the description at sixty feet. These trees were sold for twenty pounds apiece.

In the centre of this grove there stood an oak, which, though shapely and tall on the whole, bulged out into a large excrescence about the middle of the stem. On this a pair of ravens had fixed their residence for such a series of years, that the oak was distinguished by the title of The Raven-tree. Many were the attempts of the neighbouring youths to get at this eyry: the difficulty whetted their inclinations, and each was ambitious of surmounting the arduous task. But, when they arrived at the swelling, it jutted out so in their way, and was so far beyond their grasp, that the most daring lads were awed, and acknowledged the undertaking to be too hazardous. So the ravens built on, nest upon nest, in perfect security, till the fatal day arrived in which the wood was to be levelled. It was in the month of February, when those birds usually sit. The saw was applied to the butt, the wedges were inserted into the opening, the woods echoed to the heavy blows of the beetle or mallet, the tree nodded to its fall; but still the dam sat on. At last, when it gave way, the bird was flung from her nest; and, though her parental affection deserved a better fate, was whipped down by the twigs, which brought her dead to the ground.

Letter 3

THE fossil-shells of this district, and sorts of stone, such as have fallen within my observation, must not be passed over in silence. And first I must mention, as a great curiosity, a specimen that was plowed up in the chalky fields, near the side of the Down, and given to me for the singularity of its appearance, which, to an incurious eye, seems like a petrified fish of about four inches long, the cardo passing for a head and mouth. It is in reality a bivalve of the Linnaean Genus of

Mytilus, and the species of *Crista Galli*; called by Lister, *Rastellum*; by Rumphius, *Ostreum plicatum minus*; by D'Argenville, *Auris Porci*, s. *Crista Galli*; and by those who make collections *cock's comb*. Though I applied to several such in London, I never could meet with an entire specimen; nor could I ever find in books any engraving from a perfect one. In the superb museum at Leicester-house permission was given me to examine for this article; and, though I was disappointed as to the fossil, I was highly gratified with the sight of several of the shells themselves in high preservation. This bivalve is only known to inhabit the Indian ocean, where it fixes itself to a zoophyte,* known by the name *Gorgonia*. The curious foldings of the suture the one into the other, the alternate flutings or grooves, and the curved form of my specimen being much easier expressed by the pencil than by words, I have caused it to be drawn and engraved.*

Cornua Ammonis are very common about this village. As we were cutting an inclining path* up The Hanger, the labourers found them frequently on that steep, just under the soil, in the chalk, and of a considerable size. In the lane above Well-head, in the way to Emshot, they abound in the bank in a darkish sort of marl; and are usually very small and soft: but in Clay's Pond, a little farther on, at the end of the pit, where the soil is dug out for manure, I have occasionally observed them of large dimensions, perhaps fourteen or sixteen inches in diameter. But as these did not consist of firm stone, but were formed of a kind of *terra lapidosa*, or hardened clay, as soon as they were exposed to the rains and frost they mouldered away. These seemed as if they were a very recent production.* In the chalk-pit, at the north-west end of The Hanger, large *nautili* are sometimes observed.

In the very thickest strata of our freestone, and at considerable depths, well-diggers often find large scallops or pectines, having both shells deeply striated, and ridged and furrowed alternately. They are highly impregnated with, if not wholly composed of, the stone of the quarry.

Letter 4

As in a former letter the freestone of this place has been only mentioned incidentally, I shall here become more particular.

This stone is in great request for hearth-stones, and the beds of ovens: and in lining of lime-kilns it turns to good account; for the workmen use sandy loam instead of mortar; the sand of which fluxes,[3] and runs by the intense heat, and so cases over the whole face of the kiln with a strong vitrified coat like glass that it is well preserved from injuries of weather, and endures thirty or forty years. When chiseled smooth, it makes elegant fronts for houses, equal in colour and grain to the Bath stone; and superior in one respect, that, when seasoned, it does not scale. Decent chimney-pieces are worked from it of much closer and finer grain than Portland; and rooms are floored with it; but it proves rather too soft for this purpose. It is a freestone, cutting in all directions; yet has something of a grain parallel with the horizon, and therefore should not be surbedded, but laid in the same position that it grows in the quarry.[4] On the ground abroad this firestone will not succeed for pavements, because, probably some degree of saltness prevailing within it, the rain tears the slabs to pieces.[5] Though this stone is too hard to be acted on by vinegar; yet both the white part, and even the blue rag, ferments strongly in mineral acids. Though the white stone will not bear wet, yet in every quarry at intervals there are thin strata of blue rag, which resist rain and frost; and are excellent for pitching of stables, paths and courts, and for building of dry walls against banks; a valuable species of fencing, much in use in this village, and for mending of roads. This rag is rugged and stubborn, and will not hew to a smooth face; but is very durable: yet, as these strata are shallow and lie deep, large quantities cannot be procured but at considerable expense. Among the blue rags turn up some blocks tinged with a stain of yellow or rust colour, which seem to be nearly as lasting as the blue; and every now and then balls of a friable substance, like rust of iron, called rust balls.

In Wolmer Forest I see but one sort of stone, called by the workmen sand, or forest-stone.* This is generally of the colour of rusty iron, and might probably be worked as iron ore; is very hard and heavy, and of a firm, compact texture, and composed of a small roundish crystalline grit, cemented together by a brown, terrene, ferruginous matter; will not cut without difficulty, nor easily strike fire with steel. Being often found in broad flat pieces, it makes good pavement for paths about houses, never becoming slippery in frost or rain; is excellent for dry walls, and is sometimes used in buildings. In many parts of that waste it lies scattered on the surface of the ground; but is dug on Weaver's Down, a vast hill on the eastern verge of that forest, where the pits are shallow, and the stratum thin. This stone is imperishable.

From a notion of rendering their work the more elegant, and giving it a finish, masons chip this stone into small fragments about the size of the head of a large nail; and then stick the pieces into the wet mortar along the joints of their freestone walls: this embellishment carries an odd appearance, and has occasioned strangers sometimes to ask us pleasantly, 'whether we fastened our walls together with tenpenny nails?'

Letter 5

AMONG the singularities of this place the two rocky hollow lanes, the one to Alton, and the other to the forest, deserve our attention. These roads, running through the malm lands, are, by the traffick of ages, and the fretting of water, worn down through the first stratum of our freestone, and partly through the second; so that they look more like water-courses than roads; and are bedded with naked rag for furlongs together. In many places they are reduced sixteen or eighteen feet beneath the level of the fields; and after floods, and in frosts, exhibit very grotesque and wild appearances, from the tangled roots that are twisted among the strata, and from the torrents rushing down their broken sides; and especially when those cascades are frozen into icicles, hanging in all the fanciful shapes of frost-work. These rugged gloomy scenes affright the ladies when they peep down into them from the paths above, and make timid horsemen shudder while they ride along them; but delight the naturalist with their various botany, and particularly with their curious filices* with which they abound.

The manor of Selborne, was it strictly looked after, with all its kindly aspects,* and all its sloping coverts, would swarm with game; even now hares, partridges, and pheasants abound; and in old days woodcocks were as plentiful. There are few quails, because they more affect open fields than enclosures; after harvest some few land-rails are seen.

The parish of Selborne, by taking in so much of the forest, is a vast district. Those who tread the bounds* are employed part of three days in the business, and are of opinion that the outline, in all its curves and indentings, does not comprise less than thirty miles.

The village stands in a sheltered spot, secured by The Hanger from the strong westerly winds. The air is soft, but rather moist from the effluvia* of so many trees; yet perfectly healthy and free from agues.

The quantity of rain that falls on it is very considerable, as may be supposed in so woody and mountainous a district. As my experience in measuring the water is but of short date, I am not qualified to give the mean quantity.[6] I only know that

	Inch.	Hund.
From May 1, 1779, to the end of the year, there fell	28	37!*
From Jan. 1, 1780, to Jan. 1, 1781	27	32
From Jan. 1, 1781, to Jan. 1, 1782	30	71
From Jan. 1, 1782, to Jan. 1, 1783	50	26!
From Jan. 1, 1783, to Jan. 1, 1784	33	71
From Jan. 1, 1784, to Jan. 1, 1785	33	80
From Jan. 1, 1785, to Jan. 1, 1786	31	55
From Jan. 1, 1786, to Jan. 1, 1787	39	57

The village of Selborne, and large hamlet of Oakhanger, with the single farms, and many scattered houses along the verge of the forest, contain upwards of six hundred and seventy inhabitants.[7] We abound with poor; many of whom are sober and industrious, and live comfortably in good stone or brick cottages, which are glazed, and have chambers above stairs: mud buildings we have none. Besides the employment from husbandry, the men work in hop gardens, of which we have many; and fell and bark timber.* In the spring and summer the women weed the corn; and enjoy a second harvest in September by hop picking. Formerly, in the dead months they availed themselves greatly by spinning wool, for making of barragons, a genteel corded stuff, much in vogue at that time for summer wear; and chiefly manufactured at Alton, a neighbouring town, by some of the people called Quakers: but from circumstances this trade is, at an end.[8] The inhabitants enjoy a good share of health and longevity; and the parish swarms with children.

Letter 6

SHOULD I omit to describe with some exactness the forest of Wolmer, of which three fifths perhaps lie in this parish, my account of Selborne would be very imperfect, as it is a district abounding with many curious productions, both animal and vegetable; and has often afforded me much entertainment both as a sportsman and as a naturalist.

The royal forest of Wolmer is a tract of land of about seven miles in length, by two and a half in breadth, running nearly from North to South, and is abutted on, to begin to the South, and so to proceed eastward, by the parishes of Greatham, Lysse, Rogate, and Trotton, in the county of Sussex; by Bramshot, Hedleigh, and Kingsley. This royalty consists entirely of sand covered with heath and fern; but is somewhat diversified with hills and dales, without having one standing tree in the whole extent. In the bottoms, where the waters stagnate, are many bogs, which formerly abounded with subterraneous trees; though Dr Plot says positively,[9] that 'there never were any fallen trees hidden in the mosses of the southern counties.' But he was mistaken: for I myself have seen cottages on the verge of this wild district, whose timbers consisted of a black hard wood, looking like oak, which the owners assured me they procured from the bogs by probing the soil with spits, or some such instruments: but the peat is so much cut out, and the moors have been so well examined, that none has been found of late.[10] Besides the oak, I have also been shewn pieces of fossil-wood of a paler colour, and softer nature, which the inhabitants called fir: but, upon a nice examination, and trial by fire, I could discover nothing resinous in them; and therefore rather suppose that they were parts of a willow or alder, or some such aquatic tree.*

This lonely domain is a very agreeable haunt for many sorts of wild fowls, which not only frequent it in the winter, but breed there in the summer; such as lapwings, snipes, wild-ducks, and, as I have discovered within these few years, teals.* Partridges in vast plenty are bred in good seasons on the verge of this forest, into which they love to make excursions: and in particular, in the dry summer of 1740 and 1741, and some years after, they swarmed to such a degree that parties of unreasonable* sportsmen killed twenty and sometimes thirty brace in a day.

But there was a nobler species of game in this forest, now extinct, which I have heard old people say abounded much before shooting flying* became so common, and that was the heath-cock, black game, or grouse. When I was a little boy I recollect one coming now and then to my father's table. The last pack remembered was killed about thirty-five years ago; and within these ten years one solitary grey hen was sprung by some beagles, in beating for a hare. The sportsman cried out, 'A hen pheasant'; but a gentleman present, who had often seen grouse in the north of England, assured me that it was a greyhen.

Nor does the loss of our black game prove the only gap in the *Fauna Selborniensis*; for another beautiful link in the chain of beings is wanting,* I mean the red deer, which toward the beginning of this century amounted to about five hundred head, and made a stately appearance. There is an old keeper, now alive, named Adams, whose great grandfather (mentioned in a perambulation taken in 1635), grandfather, father and self, enjoyed the head keepership of Wolmer forest in succession for more than an hundred years. This person assures me, that his father has often told him, that Queen Anne, as she was journeying on the Portsmouth road, did not think the

forest of Wolmer beneath her royal regard. For she came out
of the great road at Lippock, which is just by, and, reposing
herself on a bank smoothed for that purpose, lying about half
a mile to the east of Wolmer-pond, and still called Queen's-
bank, saw with great complacency and satisfaction the whole
herd of red deer brought by the keepers along the vale before
her, consisting then of about five hundred head. A sight this
worthy the attention of the greatest sovereign! But he further
adds that, by means of the Waltham blacks,* or, to use his
own expression, as soon as they began *blacking*, they were
reduced to about fifty head, and so continued decreasing till
the time of the late Duke of Cumberland. It is now more than
thirty years ago that his highness sent down an huntsman, and
six yeoman-prickers, in scarlet jackets laced with gold,
attended by the stag-hounds; ordering them to take every deer
in this forest alive, and to convey them in carts to Windsor. In
the course of the summer they caught every stag, some of
which showed extraordinary diversion: but, in the following
winter, when the hinds were also carried off, such fine chases
were exhibited as served the country people for matter of talk
and wonder for years afterwards. I saw myself one of the
yeoman-prickers single out a stag from the herd, and must
confess that it was the most curious feat of activity I ever
beheld, superior to any thing in Mr Astley's riding-school.*
The exertions made by the horse and deer much exceeded all
my expectations; though the former greatly excelled the latter
in speed. When the devoted* deer was separated from his
companions, they gave him, by their watches, *law*, as they
called it, for twenty minutes; when, sounding their horns, the
stop-dogs were permitted to pursue, and a most gallant scene
ensued.

Letter 7

THOUGH large herds of deer do much harm to the neighbourhood, yet the injury to the morals of the people is of more
moment than the loss of their crops. The temptation is

irresistible; for most men are sportsmen by constitution: and there is such an inherent spirit for hunting in human nature, as scarce any inhibitions can restrain. Hence, towards the beginning of this century all this country was wild about deer-stealing. Unless he was a hunter, as they affected to call themselves, no young person was allowed to be possessed of manhood or gallantry. The Waltham blacks at length committed such enormities, that government was forced to interfere with that severe and sanguinary act called the *black act*,[11] which now comprehends more felonies than any law that ever was framed before. And, therefore, a late bishop of Winchester, when urged to re-stock Waltham-chase,[12] refused, from a motive worthy of a prelate, replying that 'it had done mischief enough already'.

Our old race of deer-stealers are hardly extinct yet: it was but a little while ago that, over their ale, they used to recount the exploits of their youth; such as watching the pregnant hind to her lair, and, when the calf was dropped, paring its feet with a penknife to the quick to prevent its escape, till it was large and fat enough to be killed; the shooting at one of their neighbours with a bullet in a turnip-field by moonshine, mistaking him for a deer; and the losing a dog in the following extraordinary manner:—Some fellows, suspecting that a calf new-fallen was deposited in a certain spot of thick fern, went, with a lurcher, to surprise it; when the parent-hind rushed out of the brake, and, taking a vast spring with all her feet close together, pitched upon the neck of the dog, and broke it short in two.

Another temptation to idleness and sporting was a number of rabbits, which possessed all the hillocks and dry places: but these being inconvenient to the huntsmen, on account of their burrows, when they came to take away the deer, they permitted the country people to destroy them all.

Such forests and wastes, when their allurements to irregularities are removed, are of considerable service to neighbourhoods that verge upon them, by furnishing them with peat and turf for their firing; with fuel for the burning their lime; and with ashes for their grasses;* and by maintaining their geese and their stock of young cattle at little or no expense.

The manor-farm of the parish of Greatham has an admitted claim, I see (by an old record taken from the Tower of London), of turning all live stock on the forest, at proper seasons, *bidentibus exceptis* [except sheep].[13] The reason, I presume, why sheep[14] are excluded, is, because, being such close grazers, they would pick out all the finest grasses, and hinder the deer from thriving.

Though by statute 4 and 5 W and Mary c. 23. 'to burn on any waste, between Candlemas and Midsummer, any grig, ling, heath and furze, goss or fern, is punishable with whipping and confinement in the house of correction'; yet, in this forest, about March or April, according to the dryness of the season, such vast heath-fires are lighted up, that they often get to a masterless head, and, catching the hedges, have sometimes been communicated to the underwoods, woods, and coppices, where great damage has ensued. The plea for these burnings is, that, when the old coat of heath, &c. is consumed, young will sprout up, and afford much tender brouze* for cattle; but, where there is large old furze, the fire, following the roots, consumes the very ground; so that for hundreds of acres nothing is to be seen but smother and desolation, the whole circuit round looking like the cinders of a volcano; and, the soil being quite exhausted, no traces of vegetation are to be found for years. These conflagrations, as they take place usually with a north-east or east wind, much annoy this village with their smoke, and often alarm the country; and, once in particular, I remember that a gentleman,* who lives beyond Andover, coming to my house, when he got on the downs between that town and Winchester, at twenty-five miles distance, was surprised much with smoke and a hot smell of fire; and concluded that Alresford was in flames; but, when he came to that town, he then had apprehensions for the next village, and so on to the end of his journey.

On two of the most conspicuous eminences of this forest stand two arbours or bowers, made of the boughs of oaks; the one called Waldon-lodge, the other Brimstone-lodge: these the keepers renew annually on the feast of St Barnabas,* taking the old materials for a perquisite. The farm called Blackmoor, in this parish, is obliged to find the posts and brush-wood for the former; while the farms at Greatham, in rotation, furnish for the latter; and are all enjoined to cut and deliver the materials at the spot. This custom I mention, because I look upon it to be of very remote antiquity.

Letter 8

On the very verge of the forest, as it is now circumscribed, are three considerable lakes, two in Oakhanger, of which I have nothing particular to say; and one called Bin's, or Bean's pond, which is worthy the attention of a naturalist or a sportsman. For, being crowded at the upper end with willows, and with the *carex cespitosa*,[15] it affords such a safe and pleasant shelter to wild ducks, teals, snipes, &c. that they breed there. In the winter this covert is also frequented by foxes, and sometimes by pheasants; and the bogs produce many curious plants.[16]

By a perambulation of Wolmer forest and The Holt, made in 1635, and the eleventh year of Charles the First (which now lies before me), it appears that the limits of the former are much circumscribed.[17] For, to say nothing of the farther side, with which I am not so well acquainted, the bounds on this side, in old times, came into Binswood; and extended to the ditch of Ward le ham-park, in which stands the curious mount, called King John's Hill, and Lodge Hill; and to the verge of Hartley Mauduit, called Mauduit-hatch; comprehending also Short-heath, Oakhanger, and Oakwoods; a large district, now private property, though once belonging to the royal domain.

It is remarkable that the term *purlieu* is never once mentioned in this long roll of parchment. It contains, besides the perambulation, a rough estimate of the value of the timbers, which were considerable, growing at that time in the district of The Holt; and enumerates the officers, superior and inferior, of those joint forests, for the time being, and their ostensible fees and perquisites. In those days, as at present, there were hardly any trees in Wolmer forest.

Within the present limits of the forest are three considerable lakes, Hogmer, Cranmer, and Wolmer; all of which are stocked with carp, tench, eels, and perch: but the fish do not thrive well, because the water is hungry, and the bottoms are a naked sand.

A circumstance respecting these ponds, though by no means peculiar to them, I cannot pass over in silence; and that is, that instinct by which in summer all the kine, whether oxen, cows, calves, or heifers, retire constantly to the water during the hotter hours; where, being more exempt from flies, and inhaling the coolness of that element, some belly deep, and some only to mid-leg, they ruminate and solace themselves from about ten in the morning till four in the afternoon, and then return to their feeding. During this great proportion of the day they drop much dung, in which insects nestle; and so supply food for the fish, which would be poorly subsisted but for this contingency. Thus Nature, who is a great economist, converts the recreation of one animal to the support of another! Thomson, who was a nice observer of natural occurrences, did not let this pleasing circumstance escape him. He says, in his *Summer*,

> A various group the herds and flocks compose:
> ⋅ ⋅ ⋅ ⋅ ⋅ on the grassy bank
> Some ruminating lie; while others stand
> Half in the flood, and, often bending, sip
> The circling surface.*

Wolmer-pond, so called, I suppose, for eminence sake,* is a vast lake for this part of the world, containing, in its whole circumference, 2,646 yards, or very near a mile and a half. The length of the north-west and opposite side is about 704 yards, and the breadth of the south-west end about 456 yards. This measurement, which I caused to be made with good exactness, gives an area of about sixty-six acres, exclusive of a large irregular arm at the north-east corner, which we did not take into the reckoning.

On the face of this expanse of waters, and perfectly secure from fowlers, lie all day long, in the winter season, vast flocks of ducks, teals, and widgeons, of various denominations;* where they preen and solace, and rest themselves, till towards sun-set, when they issue forth in little parties (for in their natural state they are all birds of the night) to feed in the brooks and meadows; returning again with the dawn of the morning. Had this lake an arm or two more, and were it planted round with thick covert (for now it is perfectly naked), it might make a valuable decoy.

Yet neither its extent, nor the clearness of its water, nor the resort of various and curious fowls, nor its picturesque* groups of cattle, can render this meer so remarkable, as the great quantity of coins that were found in its bed about forty years ago. But, as such discoveries more properly belong to the antiquities of this place, I shall suppress all particulars for the present, till I enter professedly on my series of letters respecting the more remote history of this village and district.*

Letter 9

BY way of supplement, I shall trouble you once more on this subject, to inform you that Wolmer, with her sister forest Ayles Holt, alias Alice Holt,[18] as it is called in old records, is held by grant from the crown for a term of years.

The grantees that the author remembers, are Brigadier-General Emanuel Scroope Howe, and his lady, Ruperta, who was a natural daughter of Prince Rupert by Margaret Hughs; a Mr Mordaunt, of the Peterborough family, who married a dowager Lady Pembroke; Henry Bilson Legge and lady; and now Lord Stawel, their son.

The lady of General Howe lived to an advanced age, long surviving her husband; and, at her death, left behind her many curious pieces of mechanism of her father's constructing, who was a distinguished mechanic and artist,[19] as well as warrior: and, among the rest, a very complicated clock, lately in possession of Mr Elmer,* the celebrated game-painter at Farnham, in the county of Surrey.

Though these two forests are only parted by a narrow range of enclosures, yet no two soils can be more different: for The Holt consists of a strong loam, of a miry nature, carrying a good turf, and abounding with oaks that grow to be large timber; while Wolmer is nothing but a hungry, sandy, barren waste.

The former, being all in the parish of Binsted, is about two miles in extent from north to south, and near as much from east to west, and contains within it many woodlands and lawns, and the great lodge where the grantees reside; and a smaller lodge called Goose-green; and is abutted on by the parishes of Kingsley, Frinsham, Farnham, and Bentley; all of which have right of common.

One thing is remarkable; that, though The Holt has been of old well stocked with fallow-deer, unrestrained by any pales or fences more than a common hedge, yet they were never seen within the limits of Wolmer; nor were the red deer of Wolmer ever known to haunt the thickets or glades of The Holt.

At present the deer of The Holt are much thinned and reduced by the night-hunters, who perpetually harass them in spite of the efforts of numerous keepers, and the severe penalties that have been put in force against them as often as they have been detected, and rendered liable to the lash of the law. Neither fines nor imprisonments can deter them: so impossible is it to extinguish the spirit of sporting, which seems to be inherent in human nature.

General Howe turned out some German wild boars and sows in his forests, to the great terror of the neighbourhood; and, at one time, a wild bull or buffalo: but the country rose upon them and destroyed them.

A very large fall of timber, consisting of about one thousand oaks, has been cut this spring (viz. 1784*) in The Holt forest; one fifth of which, it is said, belongs to the grantee, Lord Stawel. He lays claim also to the lop and top: but the poor of the parishes of Binsted and Frinsham, Bentley and Kingsley, assert that it belongs to them; and, assembling in a riotous manner, have actually taken it all away. One man, who keeps a team, has carried home, for his share, forty stacks of wood. Forty-five of these people his Lordship has served with actions. These trees, which were very sound, and in high perfection, were winter-cut, viz. in February and March, before the bark would run.* In old times The Holt was estimated to be eighteen miles, computed measure, from water-carriage, viz. from the town of Chertsey, on the Thames; but now it is not half that distance, since the Wey is made navigable up to the town of Godalming in the county of Surrey.

Letter 10

Aug. 4, 1767.*

I T has been my misfortune never to have had any neighbours whose studies have led them towards the pursuit of natural knowledge: so that, for want of a companion to quicken my industry and sharpen my attention, I have made but slender progress in a kind of information to which I have been attached from my childhood.*

As to swallows (*hirundines rusticae*) being found in a torpid state during the winter in the isle of Wight, or any part of this country, I never heard any such account worth attending to. But a clergyman, of an inquisitive turn, assures me, that, when he was a great boy, some workmen, in pulling down the battlements of a church tower early in the spring, found two or three swifts (*hirundines apodes*) among the rubbish, which were, at first appearance, dead; but, on being carried toward the fire, revived. He told me that, out of his great care to preserve them, he put them in a paper-bag, and hung them by the kitchen fire, where they were suffocated.

Another intelligent person has informed me that, while he was a schoolboy at Brighthelmstone, in Sussex,* a great fragment of the chalk-cliff fell down one stormy winter on the beach; and that many people found swallows among the rubbish: but, on my questioning him whether he saw any of those birds himself; to my no small disappointment, he answered me in the negative; but that others assured him they did.

Young broods of swallows began to appear this year on July the eleventh, and young martins (*hirundines urbicae*) were then fledged in their nests. Both species will breed again once. For I see by my *fauna* of last year, that young broods came forth so late as September the eighteenth. Are not these late hatchings more in favour of hiding than migration? Nay, some young martins remained in their nests last year so late as September the twenty-ninth; and yet they totally disappeared with us by the fifth of October.

How strange is it that the swift, which seems to live exactly the same life with the swallow and house-martin, should leave us before the middle of August invariably! while the latter stay often till the middle of October; and once I saw numbers of house-martins on the seventh of November. The martins and red-wing fieldfares* were flying in sight together; an uncommon assemblage of summer and winter-birds!

A little yellow bird (it is either a species of the *alauda trivialis*, or rather perhaps of the *motacilla trochilus*) still continues to make a sibilous shivering noise in the tops of tall woods.* The *stoparola* of Ray (for which we have as yet no name in these parts) is called, in your *Zoology*, the fly-catcher. There is one circumstance characteristic of this bird, which seems to have escaped observation; and that is, it takes its stand on the top of some stake or post, from whence it springs forth on its prey, catching a fly in the air, and hardly ever touching the ground, but returning still to the same stand for many times together.*

I perceive there are more than one species of the *motacilla trochilus*: Mr Derham supposes, in Ray's *Philos. Letters*, that he has discovered three. In these there is again an instance of some very common birds that have as yet no English name.

Mr Stillingfleet makes a question whether the black-cap (*motacilla atricapilla*) be a bird of passage or not: I think there is no doubt of it: for, in April, in the first fine weather, they come trooping, all at once, in these parts, but are never seen in the winter. They are delicate songsters.

Numbers of snipes breed every summer in some moory ground on the verge of this parish. It is very amusing to see the cock bird on wing at that time, and to hear his piping and humming notes.

I have had no opportunity yet of procuring any of those mice which I mentioned to you in town.* The person that brought me the last says they are plenty in harvest, at which time I will take care to get more; and will endeavour to put the matter out of doubt, whether it be a non-descript* species or not.

I suspect much there may be two species of water-rats. Ray says, and Linnaeus after him, that the water-rat is web-footed behind. Now, I have discovered a rat on the banks of our little stream that is not web-footed, and yet is an excellent swimmer and diver: it answers exactly to the *mus amphibius* of Linnaeus (see *Syst. Nat.*) which, he says, 'natat in fossis & urinatur [swims in ditches, and dives].' I should be glad to procure one 'plantis palmatis [with webbed feet].' Linnaeus seems to be in a puzzle about his *mus amphibius*, and to doubt whether it differs from his *mus terrestris*; which if it be, as he allows, the '*mus agrestis capite grandi brachyuros*' of Ray, is widely different from the water-rat, both in size, make, and manner of life.

As to the *falco*, which I mentioned in town, I shall take the liberty to send it down to you into Wales; presuming on your candour, that you will excuse me if it should appear as familiar to you as it is strange to me. Though mutilated, 'qualem dices . . . antehac fuisse, tales cum sint reliquiae! [you will tell from the remains what it was like originally!]'.

It haunted a marshy piece of ground in quest of wild-ducks and snipes: but, when it was shot, had just knocked down a rook, which it was tearing in pieces. I cannot make it answer to any of our English hawks; neither could I find any like it at the curious exhibition of stuffed birds in Spring-Gardens.* I found it nailed up at the end of a barn, which is the countryman's museum.

The parish I live in is a very abrupt, uneven country,* full of hills and woods, and therefore full of birds.

Letter 11

Selborne, Sept. 9, 1767.

It will not be without impatience that I shall wait for your thoughts with regard to the *falco*; as to its weight, breadth, &c., I wish I had set them down at the time: but, to the best of my remembrance, it weighed two pounds and eight ounces,

Hoopoe

and measured, from wing to wing, thirty-eight inches. Its cere and feet were yellow, and the circle of its eyelids a bright yellow. As it had been killed some days, and the eyes were sunk, I could make no good observation on the colour of the pupils and the irides.

The most unusual birds I ever observed in these parts were a pair of hoopoes (*upupa*), which came several years ago in the summer, and frequented an ornamented piece of ground, which joins to my garden, for some weeks. They used to march about in a stately manner, feeding in the walks, many times in the day; and seemed disposed to breed in my outlet;* but were frighted and persecuted by idle boys, who would never let them be at rest.

Three gross-beaks (*loxia coccothraustes*) appeared some years ago in my fields, in the winter; one of which I shot: since that, now and then one is occasionally seen in the same dead season.

A cross-bill (*loxia curvirostra*) was killed last year in this neighbourhood.

Our streams, which are small, and rise only at the end of the village, yield nothing but the bull's-head or miller's-thumb (*gobius fluviatilis capitatus*), the trout (*trutta fluviatilis*), the eel (*anguilla*), the lampern (*lampaetra parva et fluviatilis*), and the stickle-back (*pisciculus aculeatus*).

We are twenty miles from the sea, and almost as many from a great river, and therefore see but little of sea-birds. As to wild fowls, we have a few teams of ducks bred in the moors where the snipes breed; and multitudes of widgeons and teals in hard weather frequent our lakes in the forest.

Having some acquaintance with a tame brown owl, I find that it casts up the fur of mice, and the feathers of birds in pellets, after the manner of hawks: when full, like a dog, it hides what it cannot eat.

The young of the barn-owl are not easily raised, as they want a constant supply of fresh mice: whereas the young of the brown owl will eat indiscriminately all that is brought; snails, rats, kittens, puppies, magpies, and any kind of carrion or offal.

The house-martins have eggs still, and squab-young.* The last swift I observed was about the twenty-first of August; it was a straggler.

Red-starts, fly-catchers, white-throats, and *reguli non cristati*, still appear; but I have seen no black-caps lately.

I forgot to mention that I once saw, in Christ Church college quadrangle in Oxford, on a very sunny warm morning, a house martin flying about, and settling on the parapets, so late as the twentieth of November.

At present I know only two species of bats, the common *vespertilio murinus* and the *vespertilio auribus*.

I was much entertained last summer with a tame bat, which would take flies out of a person's hand. If you gave it any thing to eat, it brought its wings round before the mouth, hovering and hiding its head in the manner of birds of prey when they feed. The adroitness it showed in shearing off the wings of the flies, which were always rejected, was worthy of observation, and pleased me much. Insects seemed to be most acceptable, though it did not refuse raw flesh when offered: so that the notion, that bats go down chimneys and gnaw men's bacon, seems no improbable story. While I amused myself with this wonderful quadruped, I saw it several times confute the vulgar opinion, that bats when down on a flat surface cannot get on the wing again, by rising with great ease from the floor. It ran, I observed, with more dispatch than I was aware of; but in a most ridiculous and grotesque manner.

Bats drink on the wing, like swallows, by sipping the surface, as they play over pools and streams. They love to frequent* waters, not only for the sake of drinking, but on account of insects, which are found over them in the greatest plenty. As I was going some years ago, pretty late, in a boat from Richmond to Sunbury, on a warm summer's evening, I think I saw myriads of bats between the two places: the air swarmed with them all along the Thames, so that hundreds were in sight at a time.

I am, &c.

Letter 12

Nov. 4, 1767.

SIR,

It gave me no small satisfaction to hear that the *falco*[20] turned out an uncommon one. I must confess I should have been better pleased to have heard that I sent you a bird that you had never seen before; but that, I find, would be a difficult task.

I have procured some of the mice mentioned in my former letters, a young one and a female with young, both of which I have preserved in brandy.* From the colour, shape, size, and manner of nesting, I make no doubt but that the species is non-descript. They are much smaller, and more slender, than the *mus domesticus medius* of Ray; and have more of the squirrel or dor-mouse colour: their belly is white; a straight line along their sides divides the shades of their back and belly. They never enter into houses; are carried into ricks and barns with the sheaves; abound in harvest; and build their nests amidst the straws of the corn above the ground, and sometimes in thistles. They breed as many as eight at a litter, in a little round nest composed of the blades of grass or wheat.

One of these nests I procured this autumn, most artificially platted, and composed of the blades of wheat; perfectly round, and about the size of a cricket-ball; with the aperture so ingeniously closed, that there was no discovering to what part it belonged. It was so compact and well filled, that it would roll across the table without being discomposed, though it contained eight little mice that were naked and blind. As this nest was perfectly full, how could the dam come at her litter respectively so as to administer a teat to each? perhaps she opens different places for that purpose, adjusting them again when the business is over: but she could not possibly be contained herself in the ball with her young, which moreover would be daily increasing in bulk. This wonderful procreant cradle, an elegant instance of the efforts of instinct, was found in a wheat-field suspended in the head of a thistle.

A gentleman, curious in birds, wrote me word that his servant had shot one last January, in that severe weather, which he believed would puzzle me. I called to see it this summer, not knowing what to expect: but, the moment I took it in hand, I pronounced it the male *garrulus bohemicus* or German silk-tail, from the five peculiar crimson tags or points which it carries at the ends of five of the short remiges. It cannot, I suppose, with any propriety, be called an English bird: and yet I see, by Ray's *Philosoph. Letters*, that great flocks of them, feeding on haws, appeared in this kingdom in the winter of 1685.

The mention of haws puts me in mind that there is a total failure of that wild fruit, so conducive to the support of many of the winged nation. For the same severe weather, late in the spring, which cut off all the produce of the more tender and curious trees, destroyed also that of the more hardy and common.

Some birds, haunting with the missel-thrushes, and feeding on the berries of the yew-tree, which answered to the description of the *merula torquata*, or ring-ouzel, were lately seen in this neighbourhood. I employed some people to procure me a specimen, but without success. See Letter 20.

Query—Might not Canary birds be naturalized to this climate, provided their eggs were put, in the spring, into the nests of some of their congeners,* as goldfinches, greenfinches, &c.? Before winter perhaps they might be hardened, and able to shift for themselves.

About ten years ago I used to spend some weeks yearly at Sunbury, which is one of those pleasant villages lying on the Thames, near Hampton-court. In the autumn, I could not help being much amused with those myriads of the swallow kind which assemble in those parts. But what struck me most was, that, from the time they began to congregate, forsaking the chimnies and houses, they roosted every night in the osier-beds of the aits of that river. Now this resorting towards that element, at that season of the year, seems to give some countenance to the northern opinion (strange as it is) of their retiring under water. A Swedish naturalist* is so much persuaded of that fact, that he talks, in his calendar of *Flora*, as familiarly of the swallow's going under water in the beginning of September, as he would of his poultry going to roost a little before sunset.

An observing gentleman in London writes me word that he saw an house-martin, on the twenty-third of last October, flying in and out of its nest in the Borough.* And I myself, on the twenty-ninth of last October (as I was travelling through Oxford), saw four or five swallows hovering round and settling on the roof of the county-hospital.

Now is it likely that these poor little birds (which perhaps had not been hatched but a few weeks), should, at that late season of the year, and from so midland a county, attempt a voyage to Goree or Senegal, almost as far as the equator?[21]

I acquiesce entirely in your opinion—that, though most of the swallow kind may migrate, yet that some do stay behind and hide with us during the winter.

As to the short-winged soft-billed birds,* which come trooping in such numbers in the spring, I am at a loss even what to suspect about them. I watched them narrowly this year, and saw them abound till about Michaelmas, when they appeared no longer. Subsist they cannot openly among us, and yet elude the eyes of the inquisitive: and, as to their hiding, no man pretends to have found any of them in a torpid state in the winter. But with regard to their migration, what difficulties attend that supposition! that such feeble bad fliers (who the summer long never flit but from hedge to hedge) should be able to traverse vast seas and continents in order to enjoy milder seasons amidst the regions of Africa!

Letter 13

Selborne, Jan. 22, 1768.

SIR,

As in one of your former letters you expressed the more satisfaction from my correspondence on account of my living in the most southerly county; so now I may return the compliment, and expect to have my curiosity gratified by your living much more to the North.

For many years past I have observed that towards Christmas vast flocks of chaffinches have appeared in the fields; many more, I used to think, than could be hatched in any one neighbourhood. But, when I came to observe them more narrowly, I was amazed to find that they seemed to me to be almost all hens. I communicated my suspicions to some intelligent neighbours, who, after taking pains about the matter, declared that they also thought them all mostly females; at least fifty to one. This extraordinary occurrence brought to my mind the remark of Linnaeus; that 'before winter all their hen chaffinches migrate through Holland into Italy'. Now I want to know, from some curious person in the north, whether there are any large flocks of these finches with them in the winter, and of which sex they mostly consist? For, from such intelligence, one might be able to judge whether our female flocks migrate from the other end of the island, or whether they come over to us from the continent.*

We have, in the winter, vast flocks of the common linnets; more, I think, than can be bred in any one district. These, I observe, when the spring advances, assemble on some tree in the sunshine, and join all in a gentle sort of chirping, as if they were about to break up their winter quarters and betake themselves to their proper summer homes. It is well known, at least, that the swallows and the fieldfares do congregate with a gentle twittering before they make their respective departure.

You may depend on it that the bunting, *emberiza miliaria*, does not leave this county in the winter. In January 1767 I saw several dozen of them, in the midst of a severe frost, among the bushes on the downs near Andover: in our woodland enclosed district it is a rare bird.

Wagtails, both white and yellow, are with us all the winter. Quails crowd to our southern coast, and are often killed in numbers by people that go on purpose.

Mr Stillingfleet, in his *Tracts*, says that 'if the wheatear (*oenanthe*) does not quit England, it certainly shifts places; for about harvest they are not to be found, where there was before great plenty of them'. This well accounts for the vast quantities that are caught about that time on the south downs near Lewes, where they are esteemed a delicacy. There have been shepherds, I have been credibly informed, that have made many pounds in a season by catching them in traps. And though such multitudes are taken, I never saw (and I am well acquainted with those parts*) above two or three at a time: for they are never gregarious. They may perhaps migrate in general; and, for that purpose, draw towards the coast of Sussex in autumn: but that they do not all withdraw I am sure; because I see a few stragglers in many counties, at all times of the year, especially about warrens and stone quarries.

I have no acquaintance, at present, among the gentlemen of the navy: but have written to a friend, who was a sea-chaplain in the late war, desiring him to look into his minutes, with respect to birds that settled on their rigging during their voyage up or down the Channel.* What Hasselquist says on that subject is remarkable: there were little short-winged birds frequently coming on board his ship all the way from our channel quite up to the Levant, especially before squally weather.

What you suggest, with regard to Spain, is highly probable. The winters of Andalusia are so mild, that, in all likelihood, the soft-billed birds that leave us at that season may find insects sufficient to support them there.*

Some young man, possessed of fortune, health, and leisure, should make an autumnal voyage into that kingdom; and should spend a year there, investigating the natural history of that vast country. Mr Willughby[22] passed through that kingdom on such an errand; but he seems to have skirted along in a superficial manner and an ill humour, being much disgusted at the rude dissolute manners of the people.

I have no friend left now at Sunbury to apply to about the swallows roosting on the aits of the Thames: nor can I hear any more about those birds which I suspected were *merulae torquatae*.

As to the small mice, I have farther to remark, that though they hang their nests for breeding up amidst the straws of the standing corn, above the ground; yet I find that, in the winter, they burrow deep in the earth, and make warm beds of grass: but their grand rendezvous seems to be in corn-ricks, into which they are carried at harvest. A neighbour housed an oat-rick lately, under the thatch of which were assembled near an hundred, most of which were taken; and some I saw. I measured them; and found that, from nose to tail, they were just two inches and a quarter, and their tails just two inches long. Two of them, in a scale, weighed down just one copper halfpenny, which is about the third of an ounce avoirdupois; so that I suppose they are the smallest quadrupeds in this island. A full-grown *mus medius domesticus* weighs, I find, one ounce lumping weight,* which is more than six times as much as the mouse above; and measures from nose to rump four inches and a quarter, and the same in its tail. We have had a very severe frost and deep snow this month. My thermometer was one day fourteen degrees and an half below the freezing point, within doors. The tender evergreens were injured pretty much.* It was very providential that the air was still, and the ground well covered with snow, else vegetation in general must have suffered prodigiously. There is reason to believe that some days were more severe than any since the year 1739–40.

I am, &c. &c.

Letter 14

Selborne, March 12, 1768.

DEAR SIR,

If some curious gentleman would procure the head of a
fallow-deer, and have it dissected, he would find it furnished
with two spiracula, or breathing-places, besides the nostrils;
probably analogous to the *puncta lachrymalia* in the human
head. When deer are thirsty they plunge their noses, like some
horses, very deep under water, while in the act of drinking,
and continue them in that situation for a considerable time:
but, to obviate any inconveniency, they can open two vents,
one at the inner corner of each eye, having a communication
with the nose.* Here seems to be an extraordinary provision
of nature worthy our attention; and which has not, that I know
of, been noticed by any naturalist. For it looks as if these
creatures would not be suffocated, though both their mouths
and nostrils were stopped. This curious formation of the head
may be of singular service to beasts of chase, by affording them
free respiration: and no doubt these additional nostrils are
thrown open when they are hard run.²³ Mr Ray observed that,
at Malta, the owners slit up the nostrils of such asses as were
hard worked: for they, being naturally strait or small, did not
admit air sufficient to serve them when they travelled, or
laboured, in that hot climate. And we know that grooms, and
gentlemen of the turf, think large nostrils necessary, and a
perfection, in hunters and running horses.

Oppian, the Greek poet, by the following line, seems to have
had some notion that stags have four spiracula:

Τετράδυμοι ῥῖνες πίσυρες πνοίῃσι δίαυλοι

Quadrisidae nares, quadruplices ad respirationem canales.

[four nostrils, four passages for breathing. *Cynogetica* 2]

Writers, copying from one another, make Aristotle say
that goats breathe at their ears; whereas he asserts just the
contrary:—'Ἀλκμαίων γὰρ οὐκ ἀληθῆ λέγει, φάμενος ἀναπνεῖν
τὰς αἶγας κατὰ τὰ ὦτά.' 'Alcmaeon does not advance what is

true, when he avers that goats breathe through their ears.'—
History of Animals, I.xi.*

Letter 15

Selborne, March 30, 1768.

DEAR SIR,

Some intelligent country people have a notion that we have, in these parts, a species of the genus *mustelinum*, besides the weasel, stoat, ferret, and polecat; a little reddish beast, not much bigger than a field mouse, but much longer, which they call a cane. This piece of intelligence can be little depended on; but farther inquiry may be made.

A gentleman in this neighbourhood had two milkwhite rooks in one nest. A booby of a carter, finding them before they were able to fly, threw them down and destroyed them, to the regret of the owner, who would have been glad to have preserved such a curiosity in his rookery. I saw the birds myself nailed against the end of a barn, and was surprised to find that their bills, legs, feet, and claws were milkwhite.

A shepherd saw, as he thought, some white larks on a down above my house this winter: were not these the *emberiza nivalis*, the snow-flake of the *Brit. Zool.*? No doubt they were.

A few years ago I saw a cock bullfinch in a cage, which had been caught in the fields after it was come to its full colours. In about a year it began to look dingy; and, blackening every succeeding year, it became coal-black at the end of four. Its chief food was hempseed. Such influence has food on the colour of animals!* The pied and mottled colours of domesticated animals are supposed to be owing to high, various, and unusual food.

I had remarked, for years, that the root of the cuckoo-pint (*arum*) was frequently scratched out of the dry banks of hedges, and in severe snowy weather. After observing, with some exactness, myself, and getting others to do the same, we found it was the thrush kind that searched it out. The root of the *arum* is remarkably warm and pungent.

Our flocks of female chaffinches have not yet forsaken us.
The blackbirds and thrushes are very much thinned down by
that fierce weather in January.

In the middle of February I discovered, in my tall hedges, a
little bird that raised my curiosity: it was of that yellow-green
colour that belongs to the *salicaria* kind, and, I think, was soft-
billed. It was no *parus*; and was too long and too big for the
golden-crowned wren, appearing most like the largest willow-
wren. It hung sometimes with its back downwards, but never
continuing one moment in the same place. I shot at it, but it
was so desultory that I missed my aim.

I wonder that the stone curlew, *charadrius oedicnemus*, should
be mentioned by the writers as a rare bird: it abounds in all
the campaign parts of Hampshire and Sussex, and breeds, I
think, all the summer, having young ones, I know, very late in
the autumn. Already they begin clamouring in the evening.
They cannot, I think, with any propriety, be called, as they
are by Mr Ray, 'circa aquas versantes [circling over the
water]'; for with us, by day at least, they haunt only the most
dry, open, upland fields and sheep walks, far removed from
water: what they may do in the night I cannot say. Worms are
their usual food, but they also eat toads and frogs.

I can shew you some good specimens of my new mice.
Linnaeus perhaps would call the species *mus minimus*.

Letter 16

Selborne, April 18, 1768.

DEAR SIR,

The history of the stone curlew, *charadrius oedicnemus*, is as
follows. It lays its eggs, usually two, never more than three, on
the bare ground, without any nest, in the field; so the
countryman, in stirring his fallows, often destroys them. The
young run immediately from the egg like partridges, &c.* and
are withdrawn to some flinty field by the dam, where they
sculk among the stones, which are their best security; for their
feathers are so exactly of the colour of our grey spotted flints,

that the most exact observer, unless he catches the eye of the young bird, may be eluded. The eggs are short and round; of a dirty white, spotted with dark bloody blotches. Though I might not be able, just when I pleased, to procure you a bird, yet I could shew you them almost any day; and any evening you may hear them round the village, for they make a clamour which may be heard a mile. *Oedicnemus* is a most apt and expressive name for them, since their legs seem swoln like those of a gouty man. After harvest I have shot them before the pointers in turnip-fields.

I make no doubt but there are three species of the willow-wrens: two I know perfectly; but have not been able yet to procure the third. No two birds can differ more in their notes, and that constantly, than those two that I am acquainted with; for the one has a joyous, easy, laughing note; the other a harsh loud chirp. The former is every way larger, and three-quarters of an inch longer, and weighs two drams and an half; while the latter weighs but two: so the songster is one fifth heavier than the chirper. The chirper (being the first summer-bird of passage that is heard, the wryneck sometimes excepted) begins his two notes in the middle of March, and continues them through the spring and summer till the end of August, as appears by my journals. The legs of the larger of these two are flesh-coloured; of the less, black.

The grasshopper-lark began his sibilous note in my fields last Saturday. Nothing can be more amusing than the whisper of this little bird, which seems to be close by though at an hundred yards distance; and, when close at your ear, is scarce any louder than when a great way off. Had I not been a little acquainted with insects, and known that the grasshopper kind is not yet hatched, I should have hardly believed but that it had been a *locusta* whispering in the bushes. The country people laugh when you tell them that it is the note of a bird. It is a most artful creature, sculking in the thickest part of a bush; and will sing at a yard distance, provided it be concealed. I was obliged to get a person to go on the other side of the hedge where it haunted; and then it would run, creeping like a mouse, before us for an hundred yards together, through the bottom of the thorns; yet it would not come into fair sight:

but in a morning early, and when undisturbed, it sings on the top of a twig, gaping and shivering with its wings. Mr Ray himself had no knowledge of this bird, but received his account from Mr Johnson,* who apparently confounds it with the *reguli non cristati*, from which it is very distinct. See Ray's *Philos. Letters*, p. 108.

The fly-catcher (*stoparola*) has not yet appeared: it usually breeds in my vine. The redstart begins to sing: its note is short and imperfect, but is continued till about the middle of June. The willow-wrens (the smaller sort) are horrid pests in a garden, destroying the pease, cherries, currants, &c.; and are so tame that a gun will not scare them.

A List of the Summer Birds of Passage discovered in this neighbourhood, ranged somewhat in the order in which they appear:

	Linnaei Nomina*
Smallest willow-wren	*Motacilla trochilus*
Wryneck	*Jynx torquilla*
House-swallow	*Hirundo rustica*
Martin	*Hirundo urbica*
Sand-martin	*Hirundo riparia*
Cuckoo	*Cuculus canorus*
Nightingale	*Motacilla luscinia*
Blackcap	*Motacilla atricapilla*
Whitethroat	*Motacilla sylvia*
Middle willow-wren	*Motacilla trochilus*
Swift	*Hirundo apus*
Stone curlew?	*Charadrius oedicnemus*
Turtle-dove?	*Turtur aldrovandi*
Grasshopper-lark	*Alauda trivialis*
Landrail	*Rallus crex*
Largest willow-wren	*Motacilla trochilus*
Redstart	*Motacilla phoenicurus*
Goatsucker, or fern-owl	*Caprimulgus europaeus*
Fly-catcher	*Muscicapa grisola*

My countrymen talk much of a bird that makes a clatter with its bill against a dead bough, or some old pales, calling it a jar-bird. I procured one to be shot in the very fact; it proved to be the *sitta europaea* (the nuthatch). Mr Ray says that the less spotted woodpecker does the same. This noise may be heard a furlong or more.

Now is the only time to ascertain the short-winged summer birds; for, when the leaf is out, there is no making any remarks on such a restless tribe; and, when once the young begin to appear, it is all confusion: there is no distinction of genus, species, or sex.

In breeding-time snipes play over the moors, piping and humming: they always hum as they are descending. Is not their hum ventriloquous like that of the turkey? Some suspect that it is made by their wings.

This morning I saw the golden-crowned wren, whose crown glitters like burnished gold. It often hangs like a titmouse, with its back downwards.

Yours, &c. &c.

Letter 17

Selborne, June 18, 1768.

DEAR SIR,

On Wednesday last arrived your agreeable letter of June the 10th. It gives me great satisfaction to find that you pursue these studies still with such vigour, and are in such forwardness with regard to reptiles and fishes.

The reptiles, few as they are, I am not acquainted with, so well as I could wish, with regard to their natural history. There is a degree of dubiousness and obscurity attending the propagation of this class of animals, something analogous to that of the *cryptogamia* in the sexual system of plants: and the case is the same with regard to some of the fishes; as the eel, &c.

The method in which toads procreate and bring forth seems to be very much in the dark. Some authors say that they are viviparous: and yet Ray classes them among his oviparous animals; and is silent with regard to the manner of their bringing forth. Perhaps they may be ἔσω μὲν ᾠοτόκοι, ἔξω δὲ ζῳοτόκοι [internally oviparous, but externally viviparous], as is known to be the case with the viper.

The copulation of frogs (or at least the appearance of it; for

Swammerdam proves that the male has no *penis intrans*) is notorious to every body; because we see them sticking upon each other's backs for a month together in the spring: and yet I never saw, or read, of toads being observed in the same situation. It is strange that the matter with regard to the venom of toads has not been yet settled. That they are not noxious to some animals is plain: for ducks, buzzards, owls, stone curlews, and snakes, eat them, to my knowledge, with impunity. And I well remember the time, but was not eye-witness to the fact (though numbers of persons were) when a quack, at this village, ate a toad to make the country-people stare; afterwards he drank oil.*

I have been informed also, from undoubted authority, that some ladies (ladies you will say of peculiar taste) took a fancy to a toad, which they nourished summer after summer, for many years, till he grew to a monstrous size, with the maggots which turn to flesh flies. The reptile used to come forth every evening from an hole under the garden-steps; and was taken up, after supper, on the table to be fed. But at last a tame raven, kenning him as he put forth his head, gave him such a severe stroke with his horny beak as put out one eye. After this accident the creature languished for some time and died.

I need not remind a gentleman of your extensive reading of the excellent account there is from Mr Derham, in Ray's *Wisdom of God in the Creation* (p. 365), concerning the migration of frogs from their breeding ponds.* In this account he at once subverts that foolish opinion of their dropping from the clouds in rain; shewing that it is from the grateful coolness and moisture of those showers that they are tempted to set out on their travels, which they defer till those fall. Frogs are as yet in their tadpole state; but, in a few weeks, our lanes, paths, fields, will swarm for a few days with myriads of those emigrants, no larger than my little finger nail. Swammerdam gives a most accurate account of the method and situation in which the male impregnates the spawn of the female. How wonderful is the economy of Providence with regard to the limbs of so vile a reptile! While it is an aquatic it has a fish-like tail, and no legs: as soon as the legs sprout, the tail drops off as useless, and the animal betakes itself to the land!

Merret, I trust, is widely mistaken when he advances that the *rana arborea* is an English reptile; it abounds in Germany and Switzerland.*

It is to be remembered that the *salamandra aquatica* of Ray (the water-newt or eft) will frequently bite at the angler's bait, and is often caught on his hook. I used to take it for granted that the *salamandra aquatica* was hatched, lived, and died, in the water. But John Ellis, Esq. F.R.S. (the coralline Ellis)* asserts, in a letter to the Royal Society, dated June the 5th, 1766, in his account of the mud iguana, an amphibious *bipes* from South Carolina, that the water-eft, or newt, is only the larva of the land-eft, as tadpoles are of frogs.* Lest I should be suspected to misunderstand his meaning, I shall give it in his own words. Speaking of the *opercula* or coverings to the gills of the mud iguana, he proceeds to say that, 'The form of these pennated coverings approach very near to what I have some time ago observed in the larva or aquatic state of our English *lacerta*, known by the name of eft, or newt; which serve them for coverings to their gills, and for fins to swim with while in this state; and which they lose, as well as the fins of their tails, when they change their state and become land animals, as I have observed, by keeping them alive for some time myself.'

Linnaeus, in his *Systema Naturae*, hints at what Mr Ellis advances more than once.

Providence has been so indulgent to us as to allow of but one venomous reptile of the serpent kind in these kingdoms, and that is the viper.* As you propose the good of mankind to be an object of your publications, you will not omit to mention common sallad-oil as a sovereign remedy against the bite of the viper. As to the blind worm (*anguis fragilis*, so called because it snaps in sunder with a small blow), I have found, on examination, that it is perfectly innocuous. A neighbouring yeoman (to whom I am indebted for some good hints) killed and opened a female viper about the twenty-seventh of May: he found her filled with a chain of eleven eggs, about the size of those of a blackbird; but none of them were advanced so far towards a state of maturity as to contain any rudiments of young. Though they are oviparous, yet they are viviparous also, hatching their young within their bellies, and then bringing them forth. Whereas snakes lay chains of eggs every summer in my melon beds, in spite of all that my people can do to prevent them; which eggs do not hatch till the spring following, as I have often experienced. Several intelligent folks assure me that they have seen the viper open her mouth and admit her helpless young down her throat on sudden surprises, just as the female opossum does her brood into the pouch under her belly, upon the like emergencies; and yet the London viper-catchers insist on it, to Mr Barrington, that no such thing ever happens. The serpent kind eat, I believe, but once in a year; or, rather, but only just at one season of the year. Country people talk much of a water-snake, but, I am pretty sure, without any reason; for the common snake (*coluber natrix*) delights much to sport in the water, perhaps with a view to procure frogs and other food.

I cannot well guess how you are to make out your twelve species of reptiles, unless it be by the various species, or rather varieties, of our *lacerti*, of which Ray enumerates five. I have not had opportunity of ascertaining these; but remember well to have seen, formerly, several beautiful green *lacerti* on the sunny sandbanks near Farnham, in Surrey; and Ray admits there are such in Ireland.

Letter 18

Selborne, July 27, 1768.

DEAR SIR,

I received your obliging and communicative letter of June the 28th, while I was on a visit at a gentleman's house,* where I had neither books to turn to, nor leisure to sit down, to return you an answer to many queries, which I wanted to resolve in the best manner that I am able.

A person, by my order, has searched our brooks, but could find no such fish as the *gasterosteus pungitius*: he found the *gasterosteus aculeatus* in plenty. This morning, in a basket, I packed a little earthern pot full of wet moss, and in it some sticklebacks, male and female; the females big with spawn:

some lamperns; some bulls heads; but I could procure no minnows. This basket will be in Fleet-street by eight this evening; so I hope Mazel* will have them fresh and fair to-morrow morning. I gave some directions, in a letter, to what particulars the engraver should be attentive.

Finding, while I was on a visit, that I was within a reasonable distance of Ambresbury, I sent a servant over to that town, and procured several living specimens of loaches, which he brought, safe and brisk, in a glass decanter. They were taken in the gullies that were cut for watering the meadows. From these fishes (which measured from two to four inches in length) I took the following description: 'The loach, in its general aspect, has a pellucid appearance: its back is mottled with irregular collections of small black dots, not reaching much below the *linea literalis*, as are the back and tail fins: a black line runs from each eye down to the nose; its belly is of a silvery white; the upper jaw projects beyond the lower, and is surrounded with six feelers, three on each side: its pectoral fins are large, its ventral much smaller; the fin behind its anus small; its dorsal-fin large, containing eight spines; its tail, where it joins to the tail-fin, remarkably broad, without any taperness, so as to be characteristic of this genus: the tail-fin is broad, and square at the end. From the breadth and muscular strength of the tail it appears to be an active nimble fish.'

In my visit I was not very far from Hungerford, and did not forget to make some inquiries concerning the wonderful method of curing cancers by means of toads. Several intelligent persons, both gentry and clergy, do, I find, give a great deal of credit to what was asserted in the papers:* and I myself dined with a clergyman who seemed to be persuaded that what is related is matter of fact; but, when I came to attend to his account, I thought I discerned circumstances which did not a little invalidate the woman's story of the manner in which she came by her skill. She says of herself 'that, labouring under a virulent cancer, she went to some church where there was a vast crowd: on going into a pew, she was accosted by a strange clergyman; who, after expressing compassion for her situation, told her that if she would make such an application of living

toads as is mentioned she would be well.' Now is it likely that
this unknown gentleman should express so much tenderness
for this single sufferer, and not feel any for the many thousands
that daily languish under this terrible disorder? Would he not
have made use of this invaluable nostrum for his own emolu-
ment; or, at least, by some means of publication or other, have
found a method of making it public for the good of mankind?
In short, this woman (as it appears to me) having set up for a
cancer-doctress, finds it expedient to amuse the country with
this dark and mysterious relation.

The water-eft has not, that I can discern, the least appearance of any gills; for want of which it is continually rising to
the surface of the water to take in fresh air. I opened a big-
bellied one indeed, and found it full of spawn. Not that this
circumstance at all invalidates the assertion that they are
larvae: for the larvae of insects are full of eggs, which they
exclude the instant they enter their last state. The water-eft is
continually climbing over the brims of the vessel, within which
we keep it in water, and wandering away: and people every
summer see numbers crawling, out of the pools where they are
hatched, up the dry banks. There are varieties of them,
differing in colour; and some have fins up their tail and back,
and some have not.

Letter 19

Selborne, Aug. 17, 1768.

DEAR SIR,

I have now, past dispute, made out three distinct species of
the willow-wrens (*motacillae trochili*), which constantly and
invariably use distinct notes.* But, at the same time, I am
obliged to confess that I know nothing of your willow-lark.[24] In
my letter of April the 18th, I had told you peremptorily that I
knew your willow-lark, but had not seen it then: but, when I
came to procure it, it proved, in all respects, a very *motacilla
trochilus*; only that it is a size larger than the two other, and the
yellow-green of the whole upper part of the body is more vivid,

and the belly of a clearer white. I have specimens of the three sorts now lying before me; and can discern that there are three gradations of sizes, and that the least has black legs, and the other two flesh-coloured ones. The yellowest bird is considerably the largest, and has its quill-feathers and secondary feathers tipped with white, which the others have not. This last haunts only the tops of trees in high beechen woods, and makes a sibilous grasshopper-like noise, now and then, at short intervals, shivering a little with its wings when it sings; and is, I make no doubt now, the *regulus non cristatus* of Ray; which he says '*cantat voce stridula locustae* [sings with the rasping sound of a grasshopper. See *Ornithology*, 1678]'. Yet this great ornithologist never suspected that there were three species.

Letter 20

Selborne, Oct. 8, 1768.

IT is, I find, in zoology as it is in botany: all nature is so full, that that district produces the greatest variety which is the most examined. Several birds, which are said to belong to the north only, are, it seems, often in the south. I have discovered this summer three species of birds with us, which writers mention as only to be seen in the northern counties. The first that was brought me (on the 14th of May), was the sandpiper, *tringa hypoleucus*: it was a cock bird, and haunted the banks of some ponds near the village; and, as it had a companion, doubtless intended to have bred near that water. Besides, the owner has told me since, that, on recollection, he has seen some of the same birds round his ponds in former summers.

The next bird that I procured (on the 21st of May) was a male red-backed butcher bird, *lanius collurio*. My neighbour, who shot it, says that it might easily have escaped his notice, had not the outcries and chattering of the white-throats and other small birds drawn his attention to the bush where it was: its craw was filled with the legs and wings of beetles.

The next rare birds (which were procured for me last week) were some ring-ousels, *turdi torquati*.

This week twelve months a gentleman from London,* being with us, was amusing himself with a gun, and found, he told us, on an old yew hedge where there were berries, some birds like blackbirds, with rings of white round their necks: a neighbouring farmer also at the same time observed the same; but, as no specimens were procured, little notice was taken. I mentioned this circumstance to you in my letter of November the 4th, 1767: (you however paid but small regard to what I said, as I had not seen these birds myself): but last week the aforesaid farmer, seeing a large flock, twenty or thirty of these birds, shot two cocks and two hens: and says, on recollection, that he remembers to have observed these birds again last spring, about Lady-day, as it were, on their return to the north. Now perhaps these ousels are not the ousels of the north of England, but belong to the more northern parts of Europe; and may retire before the excessive rigor of the frosts in those parts; and return to breed in the spring, when the cold abates. If this be the case, here is discovered a new bird of winter passage, concerning whose migrations the writers are silent: but if these birds should prove the ousels of the north of England, then here is a migration disclosed within our own kingdom never before remarked. It does not yet appear whether they retire beyond the bounds of our island to the south; but it is most probable that they usually do, or else one cannot suppose that they would have continued so long unnoticed in the southern counties.* The ousel is larger than a blackbird, and feeds on haws; but last autumn (when there were no haws) it fed on yew-berries: in the spring it feeds on ivy-berries, which ripen only at that season, in March and April.

I must not omit to tell you (as you have been so lately on the study of reptiles) that my people, every now and then of late, draw up with a bucket of water from my well, which is 63 feet deep, a large black warty lizard with a fin-tail and yellow belly. How they first came down at that depth, and how they were ever to have got out thence without help, is more than I am able to say.

My thanks are due to you for your trouble and care in the examination of a buck's head. As far as your discoveries reach at present, they seem much to corroborate my suspicions; and I hope Mr——may find reason to give his decision in my favour; and then, I think, we may advance this extraordinary provision of nature as a new instance of the wisdom of God in the creation.*

As yet I have not quite done with my history of the *oedicnemus*, or stone-curlew; for I shall desire a gentleman in Sussex (near whose house these birds congregate in vast flocks in the autumn) to observe nicely when they leave him, (if they do leave him) and when they return again in the spring: I was with this gentleman lately,* and saw several single birds.

Letter 21

Selborne, Nov. 28, 1768.

DEAR SIR,

With regard to the *oedicnemus*, or stone-curlew, I intend to write very soon to my friend near Chichester, in whose neighbourhood these birds seem most to abound; and shall urge him to take particular notice when they begin to congregate, and afterwards to watch them most narrowly whether they do not withdraw themselves during the dead of the winter. When I have obtained information with respect to this circumstance, I shall have finished my history of the stone-curlew; which I hope will prove to your satisfaction, as it will be, I trust, very near the truth. This gentleman, as he occupies a large farm of his own, and is abroad early and late, will be a very proper spy upon the motions of these birds: and besides, as I have prevailed on him to buy the *Naturalist's Journal* (with which he is much delighted), I shall expect that he will be very exact in his dates. It is very extraordinary, as you observe, that a bird so common with us should never straggle to you.

And here will be the properest place to mention, while I think of it, an anecdote which the above-mentioned gentleman told me when I was last at his house; which was that, in a warren joining to his outlet, many daws (*corvi monedulae*) build every year in the rabbit-burrows under ground. The way he and his brothers used to take their nests, while they were boys, was by listening at the mouths of the holes; and, if they heard the young ones cry, they twisted the nest out with a forked stick. Some water-fowls (viz. the puffins) breed, I know, in that manner; but I should never have suspected the daws of building in holes on the flat ground.

Another very unlikely spot is made use of by daws as a place to breed in, and that is Stonehenge. These birds deposit their nests in the interstices between the upright and the impost-stones of that amazing work of antiquity: which circumstance alone speaks the prodigious height of the upright stones, that they should be tall enough to secure those nests from the annoyance of shepherd-boys, who are always idling round that place.

One of my neighbours last Saturday, November the 26th, saw a martin in a sheltered bottom: the sun shone warm, and the bird was hawking briskly after flies. I am now perfectly satisfied that they do not all leave this island in the winter.

You judge very right, I think, in speaking with reserve and caution concerning the cures done by toads: for, let people advance what they will on such subjects, yet there is such a propensity in mankind towards deceiving and being deceived,* that one cannot safely relate any thing from common report, especially in print, without expressing some degree of doubt and suspicion.

Your approbation, with regard to my new discovery of the migration of the ring-ousel, gives me satisfaction; and I find you concur with me in suspecting that they are foreign birds which visit us. You will be sure, I hope, not to omit to make inquiry whether your ring-ousels leave your rocks in the autumn. What puzzles me most, is the very short stay they make with us; for in about three weeks they are all gone. I shall be very curious to remark whether they will call on us at their return in the spring, as they did last year.

I want to be better informed with regard to ichthyology. If fortune had settled me near the sea-side, or near some great river, my natural propensity would soon have urged me to have made myself acquainted with their productions; but as I have lived mostly in inland parts, and in an upland district, my knowledge of fishes extends little farther than to those common sorts which our brooks and lakes produce.

I am, &c.

Letter 22

Selborne, Jan. 2, 1769.

DEAR SIR,

As to the peculiarity of jackdaws building with us under the ground in rabbit-burrows, you have, in part, hit upon the reasons; for, in reality, there are hardly any towers or steeples in all this country. And perhaps, Norfolk excepted, Hampshire and Sussex are as meanly furnished with churches as almost any counties in the kingdom. We have many livings of two or three hundred pounds a year, whose houses of worship make little better appearance than dovecots. When I first saw Northamptonshire, Cambridgeshire and Huntingdonshire, and the fens of Lincolnshire, I was amazed at the number of spires which presented themselves in every point of view.* As an admirer of prospects, I have reason to lament this want in my own country; for such objects are very necessary ingredients in an elegant landscape.

What you mention with respect to reclaimed toads raises my curiosity. An ancient author, though no naturalist, has well remarked that, 'Every kind of beasts, and of birds, and of serpents, and things in the sea, is tamed and hath been tamed, of mankind.'[25]

It is a satisfaction to me to find that a green lizard has actually been procured for you in Devonshire; because it corroborates my discovery, which I made many years ago, of the same sort, on a sunny sandbank near Farnham, in Surrey. I am well acquainted with the south hams of Devonshire; and can suppose that district, from its southerly situation, to be a proper habitation for such animals in their best colours.

Since the ring-ousels of your vast mountains do certainly

Heron

not forsake them against winter, our suspicions that those which visit this neighbourhood about Michaelmas are not English birds, but driven from the more northern parts of Europe by the frosts, are still more reasonable; and it will be worth your pains to endeavour to trace from whence they come, and to inquire why they make so very short a stay.

In your account of your error with regard to the two species of herons, you incidentally gave me great entertainment in your description of the heronry at Cressi-hall; which is a curiosity I never could manage to see. Fourscore nests of such a bird on one tree is a rarity which I would ride half as many miles to have a sight of. Pray be sure to tell me in your next whose seat Cressi-hall is, and near what town it lies.[26] I have often thought that those vast extents of fens have never been sufficiently explored. If half a dozen gentlemen, furnished with a good strength of water-spaniels, were to beat them over for a week, they would certainly find more species.

There is no bird, I believe, whose manners I have studied more than that of the *caprimulgus* (the goat-sucker), as it is a wonderful and curious creature; but I have always found that though sometimes it may chatter as it flies, as I know it does, yet in general it utters its jarring note sitting on a bough; and I have for many an half hour watched it as it sat with its under mandible quivering, and particularly this summer. It perches usually on a bare twig, with its head lower than its tail, in an attitude well expressed by your draughtsman in the folio *British Zoology*.* This bird is most punctual in beginning its song exactly at the close of day; so exactly that I have known it strike up more than once or twice just at the report of the Portsmouth evening gun, which we can hear when the weather is still. It appears to me past all doubt that its notes are formed by organic impulse, by the powers of the parts of its windpipe, formed for sound, just as cats purr. You will credit me, I hope, when I assure you that, as my neighbours were assembled in an hermitage on the side of a steep hill where we drink tea,* one of these churn-owls came and settled on the cross of that little straw edifice and began to chatter, and continued his note for many minutes: and we were all struck with wonder to find that the organs of that little animal, when put in motion, gave a sensible vibration to the whole building! This bird also sometimes makes a small squeak, repeated four or five times; and I have observed that to happen when the cock has been pursuing the hen in a toying way through the boughs of a tree.

It would not be at all strange if your bat, which you have procured, should prove a new one, since five species have been found in a neighbouring kingdom. The great sort that I mentioned is certainly a non-descript: I saw but one this summer, and that I had no opportunity of taking.

Your account of the Indian-grass* was entertaining. I am no angler myself; but inquiring of those that are, what they supposed that part of their tackle to be made of? they replied 'of the intestines of a silkworm.'

Though I must not pretend to great skill in entomology, yet I cannot say that I am ignorant of that kind of knowledge: I may now and then perhaps be able to furnish you with a little information.

The vast rain ceased with us much about the same time as with you, and since we have had delicate weather. Mr Barker, who has measured the rain for more than thirty years, says, in a late letter, that more rain has fallen this year than in any he ever attended to; though, from July 1763 to January 1764, more fell than in any seven months of this year.*

Letter 23

Selborne, Feb. 28, 1769.

DEAR SIR,

It is not improbable that the Guernsey lizard and our green lizards may be specifically the same; all that I know is, that, when some years ago many Guernsey lizards were turned loose in Pembroke college garden,* in the university of Oxford, they lived a great while, and seemed to enjoy themselves very well, but never bred. Whether this circumstance will prove any thing either way I shall not pretend to say.

I return you thanks for your account of Cressi-hall; but recollect, not without regret, that in June 1746 I was visiting for a week together at Spalding, without ever being told that such a curiosity was just at hand. Pray send me word in your next what sort of tree it is that contains such a quantity of herons' nests; and whether the heronry consists of a whole grove or wood, or only a few trees.*

It gave me satisfaction to find we accorded so well about the *caprimulgus*: all I contended for was to prove that it often chatters sitting as well as flying; and therefore the noise was voluntary, and from organic impulse, and not from the resistance of the air against the hollow of its mouth and throat.

If ever I saw any thing like actual migration, it was last Michaelmas-day.* I was travelling, and out early in the morning: at first there was a vast fog; but, by the time that I was got seven or eight miles from home towards the coast, the sun broke out into a delicate warm day. We were then on a large heath or common, and I could discern, as the mist began to break away, great numbers of swallows (*hirundines rusticae*) clustering on the stunted shrubs and bushes, as if they had roosted there all night. As soon as the air became clear and pleasant they all were on the wing at once; and, by a placid and easy flight, proceeded on southward towards the sea: after this I did not see any more flocks, only now and then a straggler.

I cannot agree with those persons that assert that the swallow kind disappear some and some gradually, as they come, for the bulk of them seem to withdraw at once; only some stragglers stay behind a long while, and do never, there is the greatest reason to believe, leave this island. Swallows seem to lay themselves up, and to come forth in a warm day, as bats do continually of a warm evening, after they have disappeared for weeks. For a very respectable gentleman assured me that, as he was walking with some friends under Merton-wall on a remarkably hot noon, either in the last week in December or the first week in January, he espied three or four swallows huddled together on the moulding of one of the windows of that college. I have frequently remarked that swallows are seen later at Oxford than elsewhere: is it owing to the vast massy buildings of that place, to the many waters round it, or to what else?

When I used to rise in a morning last autumn, and see the swallows and martins clustering on the chimneys and thatch of the neighbouring cottages, I could not help being touched with a secret delight, mixed with some degree of mortification: with delight, to observe with how much ardour and punctuality those poor little birds obeyed the strong impulse towards migration, or hiding, imprinted on their minds by their great Creator; and with some degree of mortification, when I reflected that, after all our pains and inquiries, we are yet not quite certain to what regions they do migrate; and are still further embarrassed to find that some do not actually migrate at all.

These reflections made so strong an impression on my imagination, that they became productive of a composition that may perhaps amuse you for a quarter of an hour when next I have the honour of writing to you.

Letter 24

Selborne, May 29, 1769.

DEAR SIR,

The *scarabaeus fullo* I know very well, having seen it in collections; but have never been able to discover one wild in its natural state. Mr Banks told me he thought it might be found on the seacoast.*

On the thirteenth of April I went to the sheep-down, where the ring-ousels have been observed to make their appearance at spring and fall, in their way perhaps to the north or south; and was much pleased to see three birds about the usual spot. We shot a cock and a hen; they were plump and in high condition. The hen had but very small rudiments of eggs within her, which proves they are late breeders; whereas those species of the thrush kind that remain with us the whole year have fledged young before that time. In their crops was nothing very distinguishable, but somewhat that seemed like blades of vegetables nearly digested. In autumn they feed on haws and yew-berries, and in the spring on ivy-berries. I dressed one of these birds, and found it juicy and well flavoured. It is remarkable that they make but a few days' stay in their spring visit, but rest near a fortnight at Michaelmas. These birds, from the observations of three springs and two autumns, are most punctual in their return; and exhibit a new migration unnoticed by the writers, who supposed they never were to be seen in any of the southern counties.*

One of my neighbours lately brought me a new *salicaria*, which at first I suspected might have proved your willow-lark,[27] but, on a nicer examination, it answered much better to the description of that species which you shot at Revesby, in Lincolnshire. My bird I describe thus: 'It is a size less than the grasshopper-lark; the head, back, and coverts of the wings, of a dusky brown, without those dark spots of the grasshopper-lark; over each eye is a milkwhite stroke; the chin and throat are white, and the under parts of a yellowish white; the rump is tawny, and the feathers of the tail sharp-pointed; the bill is dusky and sharp, and the legs are dusky; the hinder claw long and crooked.' The person that shot it says that it sung so like a reed-sparrow that he took it for one; and that it sings all night: but this account merits farther inquiry.* For my part, I suspect it is a second sort of *locustella*, hinted at by Dr Derham in Ray's *Letters*: see p. 108. He also procured me a grasshopper-lark.

The question that you put with regard to those genera of animals that are peculiar to America, viz. how they came there, and whence? is too puzzling for me to answer; and yet so obvious as often to have struck me with wonder. If one looks into the writers on that subject, little satisfaction is to be found. Ingenious men will readily advance plausible arguments to support whatever theory they shall chuse to maintain; but then the misfortune is, every one's hypothesis is each as good as another's, since they are all founded on conjecture. The late writers of this sort, in whom may be seen all the arguments of those that have gone before, as I remember, stock America from the western coast of Africa and the south of Europe; and then break down the Isthmus that bridged over the Atlantic.* But this is making use of a violent piece of machinery: it is a difficulty worthy of the interposition of a god! 'Incredulus odi. [Unable to provide any proof, I decline to believe].'

The Naturalist's Summer-evening Walk.

equidem credo, quia sit divinitus illis
Ingenium.

[I certainly believe they possess divine
wisdom.   Virgil, *Georgics* 1]

When day declining sheds a milder gleam,
What time the may-fly[28] haunts the pool or stream;
When the still owl skims round the grassy mead,
What time the timorous hare limps forth to feed;
Then be the time to steal adown the vale,
And listen to the vagrant[29] cuckoo's tale;

To hear the clamorous curlew[30] call his mate,
Or the soft quail his tender pain relate;
To see the swallow sweep the dark'ning plain
Belated, to support her infant train;
To mark the swift in rapid giddy ring
Dash round the steeple, unsubdu'd of wing:
Amusive birds!—say where your hid retreat
When the frost rages and the tempests beat;
Whence your return, by such nice instinct led,
When spring, soft season, lifts her bloomy head?
Such baffled searches mock man's prying pride,
The GOD of NATURE is your secret guide!

While deep'ning shades obscure the face of day,
To yonder bench leaf-shelter'd let us stray,
'Till blended objects fail the swimming sight,
And all the fading landscape sinks in night;
To hear the drowsy dor come brushing by
With buzzing wing, or the shrill cricket[31] cry;
To see the feeding bat glance through the wood;
To catch the distant falling of the flood;
While o'er the cliff th'awaken'd churn-owl hung
Through the still gloom protracts his chattering song;
While high in air, and pois'd upon his wings,
Unseen, the soft enamour'd woodlark[32] sings:
These, NATURE's works, the curious mind employ,
Inspire a soothing melancholy joy:
As fancy warms, a pleasing kind of pain
Steals o'er the cheek, and thrills the creeping vein!

Each rural sight, each sound, each smell, combine;
The tinkling sheep-bell, or the breath of kine;
The new-mown hay that scents the swelling breeze,
Or cottage-chimney smoking through the trees.

The chilling night-dews fall:—away, retire;
For see, the glow-worm lights her amorous fire![33]
Thus, e'er night's veil had half obscur'd the sky,
Th' impatient damsel hung her lamp on high:

True to the signal, by love's meteor led,
Leander hasten'd to his Hero's bed.[34]

I am, &c.

Letter 25

Selborne, Aug. 30, 1769.

DEAR SIR,

It gives me satisfaction to find that my account of the ousel migration pleases you. You put a very shrewd question when you ask me how I know that their autumnal migration is southward? Was not candour and openness the very life of natural history, I should pass over this query just as a sly commentator does over a crabbed passage in a classic;* but common ingenuousness obliges me to confess, not without some degree of shame, that I only reasoned in that case from analogy. For as all other autumnal birds migrate from the northward to us, to partake of our milder winters, and return to the northward again when the rigorous cold abates, so I concluded that the ring-ousels did the same, as well as their congeners the fieldfares; and especially as ring-ousels are known to haunt cold mountainous countries: but I have good reason to suspect since that they may come to us from the westward; because I hear, from very good authority, that they breed on Dartmore; and that they forsake that wild district about the time that our visitors appear, and do not return till late in the spring.

I have taken a great deal of pains about your *salicaria* and mine, with a white stroke over its eye and a tawny rump. I have surveyed it alive and dead, and have procured several specimens; and am perfectly persuaded myself (and trust you will soon be convinced of the same) that it is no more nor less than the *passer arundinaceus minor* of Ray. This bird, by some means or other, seems to be entirely omitted in the *British Zoology*; and one reason probably was because it is so strangely classed in Ray, who ranges it among his *picis affines*. It ought no doubt to have gone among his *aviculae cauda unicolore*, and among your slender-billed small birds of the same division. Linnaeus might with great propriety have put it into his genus of *motacilla*; and the *motacilla salicaria* of his *fauna suecica* seems to come the nearest to it. It is no uncommon bird, haunting the sides of ponds and rivers where there is covert, and the reeds and sedges of moors. The country people in some places call it the sedge-bird. It sings incessantly night and day during the breeding-time, imitating the note of a sparrow, a swallow, a sky-lark; and has a strange hurrying manner in its song. My specimens correspond most minutely to the description of your fen *salicaria* shot near Revesby. Mr Ray has given an excellent characteristic of it when he says, 'Rostrum & pedes in hac avicula multo majores sunt quam pro corporis ratione [The beak and feet of this bird are very much larger in proportion to its body].' See letter May 29, 1769.

I have got you the egg of an *oedicnemus*, or stone-curlew, which was picked up in a fallow on the naked ground: there were two; but the finder inadvertently crushed one with his foot before he saw them.

When I wrote to you last year on reptiles, I wish I had not forgot to mention the faculty that snakes have of stinking *se defendendo* [in self-defence]. I knew a gentleman who kept a tame snake, which was in its person as sweet as any animal while in good humour and unalarmed; but as soon as a stranger, or a dog or cat, came in, it fell to hissing, and filled the room with such nauseous effluvia as rendered it hardly supportable. Thus the squnck, or stonck, of Ray's *Synop. Quadr.* is an innocuous and sweet animal; but, when pressed hard by dogs and men, it can eject such a most pestilent and fetid smell and excrement, that nothing can be more horrible.

A gentleman sent me lately a fine specimen of the *lanius minor cinerascens cum macula in scapulis alba, Raii*; which is a bird that, at the time of your publishing your two first volumes of *British Zoology*, I find you had not seen. You have described it well from Edwards's drawing.*

Letter 26

Selborne, Dec. 8, 1769.

DEAR SIR,

I was much gratified by your communicative letter on your return from Scotland, where you spent, I find, some considerable time, and gave yourself good room to examine the natural curiosities of that extensive kingdom, both those of the islands, as well as those of the highlands.* The usual bane of such expeditions is hurry; because men seldom allot themselves half the time they should do: but, fixing on a day for their return, post from place to place, rather as if they were on a journey that required dispatch, than as philosophers investigating the works of nature. You must have made, no doubt, many discoveries, and laid up a good fund of materials for a future edition of the *British Zoology*; and will have no reason to repent that you have bestowed so much pains on a part of Great-Britain that perhaps was never so well examined before.

It has always been matter of wonder to me that fieldfares, which are so congenerous to thrushes and blackbirds, should never chuse to breed in England: but that they should not think even the highlands cold and northerly, and sequestered enough, is a circumstance still more strange and wonderful. The ring-ousel, you find, stays in Scotland the whole year round; so that we have reason to conclude that those migrators that visit us for a short space every autumn do not come from thence.

And here, I think, will be the proper place to mention that those birds were most punctual again in their migration this autumn, appearing, as before, about the 30th of September: but their flocks were larger than common, and their stay protracted somewhat beyond the usual time. If they came to spend the whole winter with us, as some of their congeners do, and then left us, as they do, in spring, I should not be so much struck with the occurrence, since it would be similar to that of the other winter birds of passage; but when I see them for a fortnight at Michaelmas, and again for about a week in the middle of April, I am seized with wonder, and long to be informed whence these travellers come, and whither they go, since they seem to use our hills merely as an inn or baiting place.

Your account of the greater brambling, or snow-fleck, is very amusing; and strange it is that such a short-winged bird should delight in such perilous voyages over the northern ocean! Some country people in the winter time have every now and then told me that they have seen two or three white larks on our downs; but, on considering the matter, I begin to suspect that these are some stragglers of the birds we are talking of, which sometimes perhaps may rove so far to the southward.

It pleases me to find that white hares are so frequent on the Scottish mountains, and especially as you inform me that it is a distinct species; for the quadrupeds of Britain are so few, that every new species is a great acquisition.

The eagle-owl, could it be proved to belong to us, is so majestic a bird, that it would grace our fauna much. I never was informed before where wild-geese are known to breed.

You admit, I find, that I have proved your fen *salicaria* to be the lesser reed-sparrow of Ray: and I think you may be secure that I am right; for I took very particular pains to clear up that matter, and had some fair specimens; but as they were not well preserved, they are decayed already. You will, no doubt, insert it in its proper place in your next edition. Your additional plates will much improve your work.*

De Buffon, I know, has described the water shrew-mouse: but still I am pleased to find you have discovered it in Lincolnshire, for the reason I have given in the article of the white hare.

As a neighbour was lately plowing in a dry chalky field, far removed from any water, he turned out a water-rat, that was curiously laid up in an *hybernaculum* artificially formed of grass and leaves. At one end of the burrow lay above a gallon of potatoes regularly stowed, on which it was to have supported itself for the winter. But the difficulty with me is how this *amphibius mus* came to fix its winter station at such a distance from the water. Was it determined in its choice of that place by the mere accident of finding the potatoes which were planted there; or is it the constant practice of the aquatic-rat to forsake the neighbourhood of the water in the colder months?

Though I delight very little in analogous reasoning, knowing how fallacious it is with respect to natural history; yet, in the following instance, I cannot help being inclined to think it may conduce towards the explanation of a difficulty that I have mentioned before, with respect to the invariably early retreat of the *hirundo apus*, or swift, so many weeks before its congeners; and that not only with us, but also in Andalusia, where they also begin to retire about the beginning of August.

The great large bat[35] (which by the by is at present a non-descript in England, and what I have never been able yet to procure) retires or migrates very early in the summer: it also ranges very high for its food, feeding in a different region of the air; and that is the reason I never could procure one. Now this is exactly the case with the swifts; for they take their food in a more exalted region than the other species, and are very seldom seen hawking for flies near the ground, or over the surface of the water. From hence I would conclude that these hirundines, and the larger bats, are supported by some sorts of high-flying gnats, scarabs, or *phalaenae*, that are of short continuance; and that the short stay of these strangers is regulated by the defect of their food.

By my journal it appears that curlews clamoured on to October the thirty-first; since which I have not seen or heard any. Swallows were observed on to November the third.

Letter 27

Selborne, Feb. 22, 1770.

DEAR SIR,

Hedge-hogs abound in my gardens and fields. The manner in which they eat their roots of the plantain in my grass-walks is very curious: with their upper mandible, which is much longer than their lower, they bore under the plant, and so eat the root off upwards, leaving the tuft of leaves untouched. In this respect they are serviceable, as they destroy a very

Hedgehog

troublesome weed; but they deface the walks in some measure by digging little round holes. It appears, by the dung that they drop upon the turf, that beetles are no inconsiderable part of their food. In June last I procured a litter of four or five young hedge-hogs, which appeared to be about five or six days old: they, I find, like puppies, are born blind, and could not see when they came to my hands. No doubt their spines are soft and flexible at the time of their birth, or else the poor dam would have but a bad time of it in the critical moment of parturition: but it is plain that they soon harden; for these little pigs had such stiff prickles on their backs and sides as would easily have fetched blood, had they not been handled with caution. Their spines are quite white at this age; and they have little hanging ears, which I do not remember to be discernible in the old ones. They can, in part, at this age draw their skin down over their faces; but are not able to contract themselves into a ball, as they do, for the sake of defence, when full grown. The reason, I suppose is, because the curious muscle that enables the creature to roll itself up in a ball was not then arrived at its full tone and firmness. Hedge-hogs make a deep and warm *hybernaculum* with leaves and moss, in which they conceal themselves for the winter: but I never could find that they stored in any winter provision, as some quadrupeds certainly do.

I have discovered an anecdote with respect to the fieldfare (*turdus pilaris*), which I think is particular enough: this bird, though it sits on trees in the day time, and procures the greatest part of its food from white-thorn hedges; yea, moreover, builds on very high trees; as may be seen by the *fauna suecica*; yet always appears with us to roost on the ground. They are seen to come in flocks just before it is dark, and to settle and nestle among the heath on our forest. And besides, the larkers, in dragging their nets by night, frequently catch them in the wheat-stubbles; while the bat-fowlers,* who take many red-wings in the hedges, never entangle any of this species. Why these birds, in the matter of roosting, should differ from all their congeners, and from themselves also with respect to their proceedings by day, is a fact for which I am by no means able to account.

I have somewhat to inform you of concerning the moose-deer; but in general foreign animals fall seldom in my way: my little intelligence is confined to the narrow sphere of my own observations at home.

Letter 28

Selborne, March 1770.

ON Michaelmas-day 1768 I managed to get a sight of the female moose belonging to the duke of Richmond, at Goodwood; but was greatly disappointed, when I arrived at the spot, to find that it died, after having appeared in a languishing way for some time, on the morning before. However, understanding that it was not stripped, I proceeded to examine this rare quadruped: I found it in an old green-house, slung under the belly and chin by ropes, and in a standing posture; but, though it had been dead for so short a time, it was in so putrid a state that the stench was hardly supportable. The grand distinction between this deer, and any other species that I have ever met with, consisted in the strange length of its legs;

on which it was tilted up much in the manner of the birds of the *grallae* order. I measured it, as they do an horse, and found that, from the ground to the wither, it was just five feet four inches; which height answers exactly to sixteen hands, a growth that few horses arrive at: but then, with this length of legs, its neck was remarkably short, no more than twelve inches; so that, by straddling with one foot forward and the other backward, it grazed on the plain ground, with the greatest difficulty, between its legs: the ears were vast and lopping, and as long as the neck; the head was about twenty inches long, and ass-like; and had such a redundancy of upper lip as I never saw before, with huge nostrils. This lip, travellers say, is esteemed a dainty dish in North America. It is very reasonable to suppose that this creature supports itself chiefly by browsing of trees, and by wading after water plants; towards which way of livelihood the length of legs and great lip must contribute much. I have read somewhere that it delights in eating the *nymphaea*, or water-lily. From the fore-feet to the belly behind the shoulder it measured three feet and eight inches: the length of the legs before and behind consisted a great deal in the tibia, which was strangely long; but, in my haste to get out of the stench, I forgot to measure that joint exactly. Its scut seemed to be about an inch long; the colour was a grizzly black; the mane about four inches long; the fore-hoofs were upright and shapely, the hind flat and splayed. The spring before it was only two years old, so that most probably it was not then come to its growth. What a vast tall beast must a full grown stag be! I have been told some arrive at ten feet and an half! This poor creature had at first a female companion of the same species, which died the spring before. In the same garden was a young stag, or red deer, between whom and this moose it was hoped that there might have been a breed,* but their inequality of height must have always been a bar to any commerce of the amorous kind. I should have been glad to have examined the teeth, tongue, lips, hoofs, &c. minutely; but the putrefaction precluded all farther curiosity. This animal, the keeper told me, seemed to enjoy itself best in the extreme frost of the former winter. In the house they shewed me the horn of a male moose, which had no front-antlers, but only a broad palm with some snags on the edge. The noble owner of the dead moose proposed to make a skeleton of her bones.

Please to let me hear if my female moose corresponds with that you saw; and whether you think still that the American moose and European elk are the same creature.*

I am,

With the greatest esteem, &c.

Letter 29

Selborne, May 12, 1770.

DEAR SIR,

Last month we had such a series of cold turbulent weather, such a constant succession of frost, and snow, and hail, and tempest, that the regular migration or appearance of the summer birds was much interrupted. Some did not shew themselves (at least were not heard) till weeks after their usual time; as the black-cap and white-throat; and some have not been heard yet, as the grasshopper-lark and largest willow-wren. As to the fly-catcher, I have not seen it; it is indeed one of the latest, but should appear about this time: and yet, amidst all this meteorous strife and war of the elements, two swallows discovered themselves as long ago as the eleventh of April, in frost and snow; but they withdrew quickly, and were not visible again for many days. House-martins, which are always more backward than swallows, were not observed till May came in.

Among the monogamous birds several are to be found, after pairing-time, single, and of each sex: but whether this state of celibacy is matter of choice or necessity, is not so easily discoverable. When the house-sparrows deprive my martins of their nests, as soon as I cause one to be shot, the other, be it cock or hen, presently procures a mate, and so for several times following.

I have known a dove-house infested by a pair of white owls, which made great havock among the young pigeons: one of the owls was shot as soon as possible; but the survivor readily found a mate, and the mischief went on. After some time the new pair were both destroyed, and the annoyance ceased.

Another instance I remember of a sportsman, whose zeal for the increase of his game being greater than his humanity, after pairing-time he always shot the cock-bird of every couple of partridges upon his grounds; supposing that the rivalry of many males interrupted the breed: he used to say, that, though he had widowed the same hen several times, yet he found she was still provided with a fresh paramour, that did not take her away from her usual haunt.

Again; I knew a lover of setting, an old sportsman, who has often told me that soon after harvest he has frequently taken small coveys of partridges, consisting of cock-birds alone; these he pleasantly used to call old bachelors.

There is a propensity belonging to common house-cats that is very remarkable; I mean their violent fondness for fish, which appears to be their most favourite food: and yet nature in this instance seems to have planted in them an appetite that, unassisted, they know not how to gratify: for of all quadrupeds cats are the least disposed towards water; and will not, when they can avoid it, deign to wet a foot, much less to plunge into that element.

Quadrupeds that prey on fish are amphibious; such is the otter, which by nature is so well formed for diving, that it makes great havock among the inhabitants of the waters. Not supposing that we had any of those beasts in our shallow brooks, I was much pleased to see a male otter brought to me, weighing twenty-one pounds, that had been shot* on the bank of our stream below the Priory, where the rivulet divides the parish of Selborne from Harteley-wood.

Letter 30

Selborne, Aug. 1, 1770.

DEAR SIR,

The French, I think, in general are strangely prolix in their natural history. What Linnaeus says with respect to insects holds good in every other branch: 'Verbositas praesentis saeculi, calamitas artis [The undisciplined prolixity of the present age is a disaster for art].'

Pray how do you approve of Scopoli's new work? as I admire his *Entomologia*, I long to see it.

I forgot to mention in my last letter (and had not room to insert in the former) that the male moose, in rutting time, swims from island to island, in the lakes and rivers of North-America, in pursuit of the females. My friend, the chaplain, saw one killed in the water as it was on that errand in the river St Lawrence: it was a monstrous beast, he told me; but he did not take the dimensions.*

When I was last in town our friend Mr Barrington most obligingly carried me to see many curious sights. As you were then writing to him about horns, he carried me to see many strange and wonderful specimens. There is, I remember, at Lord Pembroke's, at Wilton, an horn room furnished with more than thirty different pairs; but I have not seen that house lately.

Mr Barrington shewed me many astonishing collections of stuffed and living birds from all quarters of the world. After I had studied over the latter for a time, I remarked that every species almost that came from distant regions, such as South America, the coast of Guinea, &c. were thick-billed birds of the *loxia* and *fringilla* genera; and no *motacillae*, or *muscicapae* were to be met with. When I came to consider, the reason was obvious enough; for the hard-billed birds subsist on seeds which are easily carried on board; while the soft-billed birds, which are supported by worms and insects, or, what is a *succedaneum* for them, fresh raw meat, can meet with neither in long and tedious voyages. It is from this defect of food that our collections (curious as they are) are defective,

and we are deprived of some of the most delicate and lively
genera.

I am, &c.

Letter 31

Selborne, Sept. 14, 1770.

DEAR SIR,

You saw, I find, the ring-ousels again among their native
crags; and are farther assured that they continue resident in
those cold regions the whole year. From whence then do our
ring-ousels migrate so regularly every September and make
their appearance again, as if in their return, every April? They
are more early this year than common, for some were seen at
the usual hill on the fourth of this month.

An observing Devonshire gentleman tells me that they
frequent some parts of Dartmoor, and breed there; but leave
those haunts about the end of September or beginning of
October, and return again about the end of March.

Another intelligent person assures me that they breed in
great abundance all over the Peak of Derby, and are called
there Tor-ousels; withdraw in October and November, and
return in spring. This information seems to throw some light
on my new migration.

Scopoli's[36] new work (which I have just procured) has its
merits in ascertaining many of the birds of the Tirol and
Carniola. Monographers, come from whence they may, have,
I think, fair pretence to challenge some regard and approbation from the lovers of natural history; for, as no man can
alone investigate all the works of nature, these partial writers
may, each in their department, be more accurate in their
discoveries, and freer from errors, than more general writers;
and so by degrees may pave the way to an universal correct
natural history. Not that Scopoli is so circumstantial and
attentive to the life and conversation of his birds as I could
wish: he advances some false facts; as when he says of the
hirundo urbica that 'pullos extra nidum non nutrit [it does not

feed its young outside the nest]'. This assertion I know to be wrong, from repeated observation this summer; for house-martins do feed their young flying, though it must be acknowl-edged not so commonly as the house-swallow; and the feat is done in so quick a manner as not to be perceptible to indifferent observers. He also advances some (I was going to say) improbable facts; as when he says of the woodcock that 'pullos rostro portat fugiens ab hoste [fleeing from an enemy, it carries its young in its beak]'. But candour forbids me to say absolutely that any fact is false, because I have never been witness to such a fact. I have only to remark that the long unwieldy bill of the woodcock is perhaps the worst adapted of any among the winged creation for such a feat of natural affection.*

I am, &c.

Letter 32

Selborne, Oct. 29, 1770.

DEAR SIR,

After an ineffectual search in Linnaeus, Brisson, &c. I begin to suspect that I discern my brother's *hirundo hyberna* in Scopoli's new discovered *hirundo rupestris*, p. 167. His descrip-tion of 'Supra murina, subtus albida; rectrices macula ovali alba in latere interno; pedes nudi, nigri; rostrum nigrum; remiges obscuriores quam plumae dorsales; rectrices remigibus concolores; cauda emarginata, nec forcipata [mouse-coloured above, whitish below; tail-feathers with an oval white spot on the inner edge; feet black, with no feathers; bill black; flight feathers darker in colour than the dorsal, with the tail feathers of the same tint; tail notched, not forked]'; agrees very well with the bird in question: but when he comes to advance that it is '*statura hirundinis urbicae* [the size of a house-martin]' and that 'definitio *hirundinis ripariae* Linnaei huci quoque convenit [the Linnaean definition of a sand-martin applies to this also]', he in some measure invalidates all he has said; at least he shews at once that he compares them to these species merely from memory: for I have compared the birds themselves, and find they differ widely in every circumstance of shape, size, and colour. However, as you will have a specimen, I shall be glad to hear what your judgment is in the matter.*

Whether my brother is forestalled in his non-descript or not, he will have the credit of first discovering that they spend their winters under the warm and sheltery shores of Gibraltar and Barbary.

Scopoli's characters of his ordines and genera are clean, just, and expressive, and much in the spirit of Linnaeus. These few remarks are the result of my first perusal of Scopoli's *Annus Primus*.

The bane of our science is the comparing one animal to the other by memory: for want of caution in this particular Scopoli falls into errors: he is not so full with regard to the manners of his indigenous birds as might be wished, as you justly observe: his Latin is easy, elegant, and expressive, and very superior to Kramer's.[37]

I am pleased to see that my description of the moose corresponded so well with yours.

I am, &c.

Letter 33

Selborne, Nov. 26, 1770.

DEAR SIR,

I was much pleased to see, among the collection of birds from Gibraltar, some of those short-winged English summer-birds of passage,* concerning whose departure we have made so much inquiry. Now if these birds are found in Andalusia to migrate to and from Barbary, it may easily be supposed that those that come to us may migrate back to the continent, and spend their winters in some of the warmer parts of Europe. This is certain, that many soft-billed birds that come to Gibraltar appear there only in spring and autumn, seeming to advance in pairs towards the northward, for the sake of breeding during the summer months; and retiring in parties and broods towards the south at the decline of the year: so that the rock of Gibraltar is the great rendezvous, and place of observation, from whence they take their departure each way towards Europe or Africa. It is therefore no mean discovery, I think, to find that our small short-winged summer birds of passage are to be seen spring and autumn on the very skirts of Europe; it is a presumptive proof of their emigrations.

Scopoli seems to me to have found the *hirundo melba*, the great Gibraltar swift, in Tirol, without knowing it. For what is his *hirundo alpina* but the aforementioned bird in other words? Says he 'Omnia prioris (meaning the swift); sed pectus album; paulo major priore [Everything similar to the former, but the breast is white and the body a little larger]'. I do not suppose this to be a new species. It is true also of the *melba*, that 'nidificat in excelsis Alpium rupibus [it nests on lofty peaks in the Alps]'. Vid. *Annum Primum*.

My Sussex friend, a man of observation and good sense, but no naturalist, to whom I applied on account of the stone-curlew, *oedicnemus*, sends me the following account: 'In looking over my Naturalist's Journal for the month of April, I find the stone-curlews are first mentioned on the seventeenth and eighteenth, which date seems to me rather late. They live with us all the spring and summer, and at the beginning of autumn prepare to take leave by getting together in flocks. They seem to me a bird of passage that may travel into some dry hilly country south of us, probably Spain, because of the abundance of sheep-walks in that country; for they spend their summers with us in such districts. This conjecture I hazard, as I have never met with any one that has seen them in England in the winter. I believe they are not fond of going near the water, but feed on earth-worms, that are common on sheep-walks and downs. They breed on fallows and lay-fields abounding with grey mossy flints, which much resemble their young in colour, among which they skulk and conceal themselves. They make no nest, but lay their eggs on the bare ground, producing in common but two at a time. There is reason to think their young run soon after they are hatched; and that the old ones do not feed them, but only lead them about at the time of feeding, which, for the most part, is in the night.' Thus far my friend.

In the manners of this bird you see there is something very analogous to the bustard, whom it also somewhat resembles in aspect and make, and in the structure of its feet.

For a long time I have desired my relation to look out for these birds in Andalusia; and now he writes me word that, for the first time, he saw one dead in the market on the third of September.

When the *oedicnemus* flies it stretches out its legs straight behind, like an heron.

I am, &c.

Letter 34

Selborne, March 30, 1771.

DEAR SIR,

There is an insect with us, especially on chalky districts, which is very troublesome and teasing all the latter end of the summer, getting into people's skins, especially those of women and children, and raising tumours which itch intolerably. This animal (which we call an harvest bug) is very minute, scarce discernible to the naked eye; of a bright scarlet colour, and of the genus of *Acarus*. They are to be met with in gardens, on kidneybeans, or any legumens; but prevail only in the hot months of summer. Warreners, as some have assured me, are much infested by them on chalky downs; where these insects swarm sometimes to so infinite a degree as to discolour their nets, and to give them a reddish cast, while the men are so bitten as to be thrown into fevers.

There is a small long shining fly in these parts very troublesome to the housewife, by getting into the chimnies, and laying its eggs in the bacon while it is drying: these eggs produce maggots called *jumpers*, which, harbouring in the gammons and best part of the hogs, eat down to the bone, and make great waste. This fly I suspect to be a variety of the *musca* *putris* of Linnaeus:* it is to be seen in the summer in farm-kitchens on the bacon-racks and about the mantle-pieces, and on the ceilings.

The insect that infests turnips and many crops in the garden (destroying often whole fields while in their seedling leaves) is an animal that wants to be better known. The country people here call it the turnip-fly and black-dolphin; but I know it to be one of the *coleoptera*; the 'chrysomela oleracea, saltatoria, femoribus posticis crassissimis'. In very hot summers they abound to an amazing degree, and, as you walk in a field or in a garden, make a pattering like rain, by jumping on the leaves of the turnips or cabbages.

There is an *Oestrus*, known in these parts to every ploughboy; which, because it is omitted by Linnaeus, is also passed over by late writers; and that is the *curvicauda* of old Moufet, mentioned by Derham in his *Physico-theology*, p. 250: an insect worthy of remark for depositing its eggs as it flies in so dextrous a manner on the single hairs of the legs and flanks of grass-horses. But then Derham is mistaken when he advances that this *Oestrus* is the parent of that wonderful star-tailed maggot which he mentions afterwards; for more modern entomologists have discovered that singular production to be derived from the egg of the *musca chamaeleon*: see Geoffroy, t. 17, f. 4.*

A full history of noxious insects hurtful in the field, garden, and house, suggesting all the known and likely means of destroying them,* would be allowed by the public to be a most useful and important work. What knowledge there is of this sort lies scattered, and wants to be collected; great improvements would soon follow of course. A knowledge of the properties, economy, propagation, and in short of the life and conversation of these animals, is a necessary step to lead us to some method of preventing their depredations.

As far as I am a judge, nothing would recommend entomology more than some neat plates that should well express the generic distinctions of insects according to Linnaeus; for I am well assured that many people would study insects, could they set out with a more adequate notion of those distinctions than can be conveyed at first by words alone.

Peacock

Letter 35

Selborne, 1771.

DEAR SIR,

Happening to make a visit to my neighbour's peacocks, I could not help observing that the trains of those magnificent birds appear by no means to be their tails; those long feathers growing not from their uropygium, but all up their backs. A range of short brown stiff feathers, about six inches long, fixed in the uropygium, is the real tail, and serves as the fulcrum to prop the train, which is long and top-heavy, when set on end. When the train is up, nothing appears of the bird before but its head and neck; but this would not be the case were those long feathers fixed only in the rump, as may be seen by the turkey-cock when in a strutting attitude. By a strong muscular vibration these birds can make the shafts of their long feathers clatter like the swords of a sword-dancer; they then trample very quick with their feet, and run backwards towards the females.

I should tell you that I have got an uncommon *calculus aegogropila*, taken out of the stomach of a fat ox; it is perfectly round, and about the size of a large Seville orange;* such are, I think, usually flat.

Letter 36

Sept. 1771.

Dear Sir,

The summer through I have seen but two of that large species of bat which I call *vespertilio altivolans*, from its manner of feeding high in the air: I procured one of them, and found it to be a male; and made no doubt, as they accompanied together, that the other was a female: but, happening in an evening or two to procure the other likewise, I was somewhat disappointed, when it appeared to be also of the same sex.* This circumstance, and the great scarcity of this sort, at least in these parts, occasions some suspicions in my mind whether it is really a species, or whether it may not be the male part of the more known species, one of which may supply many females; as is known to be the case in sheep, and some other quadrupeds. But this doubt can only be cleared by a farther examination, and some attention to the sex, of more specimens: all that I know at present is, that my two were amply furnished with the parts of generation much resembling those of a boar.

In the extent of their wings they measured fourteen inches and an half; and four inches and an half from the nose to the tip of the tail: their heads were large, their nostrils bilobated, their shoulders broad and muscular; and their whole bodies fleshy and plump. Nothing could be more sleek and soft than their fur, which was of a bright chesnut colour; their maws were full of food, but so macerated that the quality could not be distinguished; their livers, kidnies, and hearts, were large, and their bowels covered with fat. They weighed each, when entire, full one ounce and one drachm. Within the ear there was somewhat of a peculiar structure that I did not understand perfectly; but refer it to the observation of the curious anatomist. These creatures sent forth a very rancid and offensive smell.

Great bat

Letter 37

Selborne, 1771.

DEAR SIR,

On the twelfth of July I had a fair opportunity of contemplating the motions of the *caprimulgus*, or fern-owl, as it was playing round a large oak that swarmed with *scarabaei solstitiales*, or fern-chafers. The powers of its wing were wonderful, exceeding, if possible, the various evolutions and quick turns of the swallow genus. But the circumstance that pleased me most was, that I saw it distinctly, more than once, put out its short leg while on the wing, and, by a bend of the head, deliver somewhat into its mouth. If it takes any part of its prey with its foot, as I have now the greatest reason to suppose it does these chafers, I no longer wonder at the use of its middle toe, which is curiously furnished with a serrated claw.

Swallows and martins, the bulk of them I mean, have forsaken us sooner this year than usual; for, on September the twenty-second, they rendezvoused in a neighbour's walnut-tree, where it seemed probable they had taken up their lodging for the night. At the dawn of the day, which was foggy, they arose all together in infinite numbers, occasioning such a rushing from the strokes of their wings against the hazy air, as might be heard to a considerable distance: since that no flock has appeared, only a few stragglers.

Some swifts staid late, till the twenty-second of August—a rare instance! for they usually withdraw within the first week.[38]

On September the twenty-fourth three or four ring-ousels appeared in my fields for the first time this season: how punctual are these visitors in their autumnal and spring migrations!*

Letter 38

Selborne, March 15, 1773.

DEAR SIR,

By my journal for last autumn it appears that the house-martins bred very late, and staid very late in these parts; for, on the first of October, I saw young martins in their nest nearly fledged;* and again, on the twenty-first of October, we had at the next house a nest full of young martins just ready to fly; and the old ones were hawking for insects with great alertness. The next morning the brood forsook their nest, and were flying round the village. From this day I never saw one of the swallow kind till November the third; when twenty, or perhaps thirty, house-martins were playing all day long by the side of the hanging wood, and over my fields. Did these small weak birds, some of which were nestlings twelve days ago, shift their quarters at this late season of the year to the other side of the northern tropic? Or rather, is it not more probable that the next church, ruin, chalk-cliff, steep covert, or perhaps sandbank, lake or pool (as a more northern naturalist would say*), may become their *hybernaculum*, and afford them a ready and obvious retreat?

We now begin to expect our vernal migration of ring-ousels every week. Persons worthy of credit assure me that ring-ousels were seen at Christmas 1770 in the forest of Bere, on the southern verge of this county. Hence we may conclude that their migrations are only internal, and not extended to the continent southward, if they do at first come at all from the northern parts of this island only, and not from the north of Europe. Come from whence they will, it is plain, from the fearless disregard that they shew for men or guns, that they have been little accustomed to places of much resort. Navigators mention that in the Isle of Ascension, and other such desolate districts, birds are so little acquainted with the human form that they settle on men's shoulders; and have no more dread of a sailor than they would have of a goat that was grazing. A young man at Lewes, in Sussex, assured me that about seven years ago ring-ousels abounded so about that town in the autumn that he killed sixteen himself in one afternoon: he added further, that some had appeared since in every autumn; but he could not find that any had been observed before the season in which he shot so many. I myself have found these birds in little parties in the autumn, cantoned all along the Sussex downs, wherever there were shrubs and bushes, from Chichester to Lewes; particularly in the autumn of 1770.

I am, &c.

Letter 39

Selborne, Nov. 9, 1773.

DEAR SIR,

As you desire me to send you such observations as may occur, I take the liberty of making the following remarks, that you may, according as you think me right or wrong, admit or reject what I here advance, in your intended new edition of the *British Zoology*.*

The osprey[39] was shot about a year ago at Frinsham-pond, a great lake, at about six miles from hence, while it was sitting on the handle of a plough and devouring a fish: it used to precipitate itself into the water, and so take its prey by surprise.

A great ash-coloured butcher-bird was shot last winter in

Tisted-park, and a red backed butcher-bird at Selborne: they are *rarae aves* in this county.

Crows go in pairs the whole year round.

Cornish choughs abound, and breed on Beachy-head and on all the cliffs of the Sussex coast.

The common wild-pigeon, or stock-dove, is a bird of passage in the south of England, seldom appearing till towards the end of November; is usually the latest winter-bird of passage. Before our beechen woods were so much destroyed we had myriads of them, reaching in strings for a mile together as they went out in a morning to feed. They leave us early in spring; where do they breed?

The people of Hampshire and Sussex call the missel-bird the storm-cock, because it sings early in the spring in blowing showery weather; its song often commences with the year: with us it builds much in orchards.

A gentleman assures me he has taken the nests of ring-ousels on Dartmoor: they build in banks on the sides of streams.

Titlarks* not only sing sweetly as they sit on trees, but also as they play and toy about on the wing; and particularly while they are descending, and sometimes as they stand on the ground.

Adanson's testimony seems to me to be a very poor evidence that European swallows migrate during our winter to Senegal: he does not talk at all like an ornithologist; and probably saw only the swallows of that country, which I know build within Governor O'Hara's hall against the roof. Had he known European swallows, would he not have mentioned the species?

The house-swallow washes by dropping into the water as it flies: this species appears commonly about a week before the house-martin, and about ten or twelve days before the swift.

In 1772 there were young house-martins in their nest till October the twenty-third.

The swift appears about ten or twelve days later than the house-swallow: viz. about the twenty-fourth or twenty-sixth of April.

Whin-chats and stone-chatters stay with us the whole year.*

Some wheat-ears continue with us the winter through.

Wagtails, all sorts, remain with us all the winter.

Bulfinches, when fed on hempseed, often become wholly black.

We have vast flocks of female chaffinches all the winter, with hardly any males among them.

When you say that in breeding-time the cock-snipes make a bleating noise, and I a drumming (perhaps I should have rather said an humming), I suspect we mean the same thing. However, while they are playing about on the wing they certainly make a loud piping with their mouths: but whether that bleating or humming is ventriloquous, or proceeds from the motion of their wings, I cannot say; but this I know, that when this noise happens the bird is always descending, and his wings are violently agitated.

Soon after the lapwings have done breeding they congregate, and leaving the moors and marshes, betake themselves to downs and sheep-walks.

Two years ago last spring the little auk was found alive and unhurt, but fluttering and unable to rise, in a lane a few miles from Alresford, where there is a great lake; it was kept awhile, but died.

I saw young teals taken alive in the ponds of Wolmer-forest in the beginning of July last, along with flappers, or young wild-ducks.

Speaking of the swift,* that page says 'its drink the dew'; whereas it should be 'it drinks on the wing'; for all the swallow kind sip their water as they sweep over the face of pools or rivers: like Virgil's bees, they drink flying: 'flumina summa libant [they sip from the surface of the stream as they fly. Virgil, *Georgics* 4]'. In this method of drinking perhaps this genus may be peculiar.

Of the sedge-bird be pleased to say it sings most part of the night; its notes are hurrying, but not unpleasing, and imitative of several birds; as the sparrow, swallow, sky-lark. When it happens to be silent in the night, by throwing a stone or clod into the bushes where it sits you immediately set it a singing; or in other words, though it slumbers sometimes, yet, as soon as it is awakened it reassumes its song.

Letter 40

Selborne, Sept. 2, 1774.

DEAR SIR,

Before your letter arrived, and of my own accord, I had been remarking and comparing the tails of the male and female swallow, and this ere any young broods appeared; so that there was no danger of confounding the dams with their *pulli*: and besides, as they were then always in pairs, and busied in the employ of nidification, there could be no room for mistaking the sexes, nor the individuals of different chimnies the one for the other. From all my observations, it constantly appeared that each sex has the long feathers in its tail that give it that forked shape; with this difference, that they are longer in the tail of the male than in that of the female.

Nightingales, when their young first come abroad, and are helpless, make a plaintive and a jarring noise; and also a snapping or cracking, pursuing people along the hedges as they walk: these last sounds seem intended for menace and defiance.

The grasshopper-lark chirps all night in the height of summer.

Swans turn white the second year, and breed the third.

Weasels prey on moles, as appears by their being sometimes caught in mole-traps.

Sparrow-hawks sometimes breed in old crows' nests, and the kestril in churches and ruins.

There are supposed to be two sorts of eels in the island of Ely. The threads sometimes discovered in eels are perhaps their young: the generation of eels is very dark and mysterious.*

Hen-harriers breed on the ground, and seem never to settle on trees.

When redstarts shake their tails they move them horizontally, as dogs do when they fawn: the tail of a wagtail, when in motion, bobs up and down like that of a jaded horse.

Hedge-sparrows have a remarkable flirt with their wings in

breeding-time: as soon as frosty mornings come they make a very piping plaintive noise.

Many birds which become silent about Midsummer reassume their notes again in September; as the thrush, blackbird, woodlark, willow-wren, &c.; hence August is by much the most mute month, the spring, summer, and autumn through. Are birds induced to sing again because the temperament of autumn resembles that of spring?

Linnaeus ranges plants geographically; palms inhabit the tropics, grasses the temperate zones, and mosses and lichens the polar circles; no doubt animals may be classed in the same manner with propriety.

House-sparrows build under eaves in the spring; as the weather becomes hotter they get out for coolness, and nest in plum-trees and apple-trees. These birds have been known sometimes to build in rooks' nests, and sometimes in the forks of boughs under rooks' nests.

As my neighbour was housing a rick he observed that his dogs devoured all the little red mice that they could catch, but rejected the common mice; and that his cats ate the common mice, refusing the red.

Red-breasts sing all through the spring, summer, and autumn. The reason that they are called autumn songsters is, because in the two first seasons their voices are drowned and lost in the general chorus; in the latter their song becomes distinguishable. Many songsters of the autumn seem to be the young cock red-breasts of that year: notwithstanding the prejudices in their favour, they do much mischief in gardens to the summer-fruits.[40]

The titmouse, which early in February begins to make two quaint notes, like the whetting of a saw, is the marsh titmouse:* the great titmouse sings with three cheerful joyous notes, and begins about the same time.

Wrens sing all the winter through, frost excepted.

House-martins came remarkably late this year both in Hampshire and Devonshire: is this cirumstance for or against either hiding or migration?

Most birds drink sipping at intervals; but pigeons take a long continued draught, like quadrupeds.

Notwithstanding what I have said in a former letter, no grey crows were ever known to breed on Dartmoor; it was my mistake.*

The appearance and flying of the *scarabaeus solstitialis*, or fern-chafer, commence with the month of July, and cease about the end of it. These scarabs are the constant food of *caprimulgi*, or fern owls, through that period. They abound on the chalky downs and in some sandy districts, but not in the clays.

In the garden of the Black-bear inn in the town of Reading is a stream or canal running under the stables and out into the fields on the other side of the road: in this water are many carps, which lie rolling about in sight, being fed by travellers, who amuse themselves by tossing them bread: but as soon as the weather grows at all severe these fishes are no longer seen, because they retire under the stables, where they remain till the return of spring. Do they lie in a torpid state? if they do not, how are they supported?

The note of the white-throat, which is continually repeated, and often attended with odd gesticulations on the wing, is harsh and displeasing. These birds seem of a pugnacious disposition; for they sing with an erected crest and attitudes of rivalry and defiance; are shy and wild in breeding-time, avoiding neighbourhoods, and haunting lonely lanes and commons; nay even the very tops of the Sussex-downs, where there are bushes and covert; but in July and August they bring their broods into gardens and orchards, and make great havock among the summer-fruits.

The black-cap has in common a full, sweet, deep, loud, and wild pipe; yet that strain is of short continuance, and his motions are desultory; but when that bird sits calmly and engages in song in earnest, he pours forth very sweet, but inward melody, and expresses great variety of soft and gentle modulations, superior perhaps to those of any of our warblers, the nightingale excepted.

Black-caps mostly haunt orchards and gardens; while they warble their throats are wonderfully distended.

The song of the redstart is superior, though somewhat like that of the white-throat: some birds have a few more notes than others. Sitting very placidly on the top of a tall tree in a village, the cock sings from morning to night: he affects neighbourhoods, and avoids solitude, and loves to build in orchards and about houses; with us he perches on the vane of a tall maypole.

The fly-catcher is of all our summer birds the most mute and the most familiar; it also appears the last of any. It builds in a vine, or a sweetbriar, against the wall of an house, or in the hole of a wall, or on the end of a beam or plate, and often close to the post of a door where people are going in and out all day long. This bird does not make the least pretension to song, but uses a little inward wailing note when it thinks its young in danger from cats or other annoyances: it breeds but once, and retires early.

Selborne parish alone can and has exhibited at times more than half the birds that are ever seen in all Sweden; the former has produced more than one hundred and twenty species, the latter only two hundred and twenty-one. Let me add also that it has shewn near half the species that were ever known in Great-Britain.[41]

On a retrospect, I observe that my long letter carries with it a quaint and magisterial air, and is very sententious; but, when I recollect that you requested stricture and anecdote, I hope you will pardon the didactic manner for the sake of the information it may happen to contain.

Letter 41

It is matter of curious inquiry to trace out how those species of soft-billed birds, that continue with us the winter through, subsist during the dead months.* The imbecility of birds seems not to be the only reason why they shun the rigour of our winters; for the robust wry-neck (so much resembling the hardy race of woodpeckers) migrates, while the feeble little golden-crowned wren, that shadow of a bird, braves our severest frosts without availing himself of houses or villages, to which most of our winter-birds crowd in distressful seasons, while this keeps aloof in fields and woods; but perhaps this may be the reason why they may often perish, and why they are almost as rare as any bird we know.

I have no reason to doubt but that the soft-billed birds, which winter with us, subsist chiefly on insects in their aurelia state. All the species of wagtails in severe weather haunt shallow streams near their spring-heads, where they never freeze; and, by wading, pick out the aurelias of the genus of *Phryganeae*, &c.[42]

Hedge-sparrows frequent sinks and gutters in hard weather, where they pick up crumbs and other sweepings: and in mild weather they procure worms, which are stirring every month in the year, as any one may see that will only be at the trouble of taking a candle to a grass-plot on any mild winter's night. Red-breasts and wrens in the winter haunt out-houses, stables, and barns, where they find spiders and flies that have laid themselves up during the cold season. But the grand support of the soft-billed birds in winter is that infinite profusion of aureliae of the *lepidoptera ordo*, which is fastened to the twigs of trees and their trunks; to the pales and walls of gardens and buildings; and is found in every cranny and cleft of rock or rubbish, and even in the ground itself.

Every species of titmouse winters with us; they have what I call a kind of intermediate bill between the hard and the soft, between the Linnaean genera of *fringilla* and *motacilla*. One species alone spends its whole time in the woods and fields, never retreating for succour in the severest seasons to houses and neighbourhoods; and that is the delicate long-tailed titmouse, which is almost as minute as the golden-crowned wren: but the blue titmouse, or nun (*parus caeruleus*), the cole-mouse (*parus ater*), the great black-headed titmouse (*fringillago*), and the marsh titmouse (*parus palustris*), all resort, at times, to buildings; and in hard weather particularly. The great titmouse, driven by stress of weather, much frequents houses; and, in deep snows, I have seen this bird, while it hung with its back downwards (to my no small delight and admiration), draw straws lengthwise from out the eaves of thatched houses, in order to pull out the flies that were

Blue Titmouse

concealed between them, and that in such numbers that they quite defaced the thatch, and gave it a ragged appearance.

The blue titmouse, or nun, is a great frequenter of houses, and a general devourer. Besides insects, it is very fond of flesh; for it frequently picks bones on dung-hills: it is a vast admirer of suet, and haunts butchers' shops. When a boy, I have known twenty in a morning caught with snap mouse-traps, baited with tallow or suet. It will also pick holes in apples left on the ground, and be well entertained with the seeds on the head of a sun-flower. The blue, marsh, and great titmice will, in very severe weather, carry away barley and oat straws from the sides of ricks.

How the wheat-ear and whin-chat support themselves in winter cannot be so easily ascertained, since they spend their time on wild heaths and warrens; the former especially, where there are stone quarries: most probably it is that their maintenance arises from the aureliae of the *lepidoptera ordo*, which furnish them with a plentiful table in the wilderness.*

I am, &c.

Letter 42

Selborne, March 9, 1775.

DEAR SIR,

Some future faunist, a man of fortune, will, I hope, extend his visits to the kingdom of Ireland; a new field, and a country little known to the naturalist. He will not, it is to be wished, undertake that tour unaccompanied by a botanist, because the mountains have scarcely been sufficiently examined; and the southerly counties of so mild an island may possibly afford some plants little to be expected within the British dominions. A person of a thinking turn of mind will draw many just remarks from the modern improvements of that country, both in arts and agriculture, where premiums obtained long before they were heard of with us. The manners of the wild natives, their superstitions, their prejudices, their sordid way of life, will extort from him many useful reflections. He should also take with him an able draughtsman; for he must by no means pass over the noble castles and seats, the extensive and picturesque lakes and waterfalls, and the lofty stupendous mountains, so little known, and so engaging to the imagination when described and exhibited in a lively manner: such a work would be well received.*

As I have seen no modern map of Scotland, I cannot pretend to say how accurate or particular any such may be; but this I know, that the best old maps of that kingdom are very defective.

The great obvious defect that I have remarked in all maps of Scotland that have fallen in my way is, a want of a coloured line, or stroke, that shall exactly define the just limits of that district called The Highlands. Moreover, all the great avenues to that mountainous and romantic country want to be well distinguished. The military roads formed by general Wade are so great and Roman-like an undertaking that they well merit attention. My old map, Moll's Map, takes notice of Fort William; but could not mention the other forts that have been erected long since: therefore a good representation of the chain of forts should not be omitted.*

The celebrated zigzag up the Coryarich must not be passed over. Moll takes notice of Hamilton and Drumlanrig, and such capital houses; but a new survey, no doubt, should represent every seat and castle remarkable for any great event, or celebrated for its paintings, &c. Lord Breadalbane's seat and beautiful policy are too curious and extraordinary to be omitted.

The seat of the Earl of Eglintoun, near Glasgow, is worthy of notice. The pine-plantations of that nobleman are very grand and extensive indeed.

I am, &c.

Letter 43

A PAIR of honey-buzzards, *buteo apivorus, sive vespivorus Raii*, built them a large shallow nest, composed of twigs and lined with dead beechen leaves, upon a tall slender beech near the middle of Selborne-hanger, in the summer of 1780. In the middle of the month of June a bold boy climbed this tree, though standing on so steep and dizzy a situation, and brought down an egg, the only one in the nest, which had been sat on for some time,* and contained the embrio of a young bird. The egg was smaller, and not so round as those of the common buzzard; was dotted at each end with small red spots, and surrounded in the middle with a broad bloody zone.

The hen-bird was shot, and answered exactly to Mr Ray's description of that species; had a black cere, short thick legs, and a long tail. When on the wing this species may be easily distinguished from the common buzzard by its hawk-like appearance, small head, wings not so blunt, and longer tail. This specimen contained in its craw some limbs of frogs and many grey snails without shells. The irides of the eyes of this bird were of a beautiful bright yellow colour.

About the tenth of July in the same summer a pair of sparrow-hawks bred in an old crow's nest on a low beech in the same hanger; and as their brood, which was numerous, began to grow up, became so daring and ravenous, that they were a terror to all the dames in the village that had chickens or ducklings under their care. A boy climbed the tree, and found the young so fledged that they all escaped from him; but discovered that a good house had been kept: the larder was well-stored with provisons; for he brought down a young blackbird, jay, and house-martin, all clean picked, and some half devoured. The old birds had been observed to make sad havock for some days among the new-flown swallows and martins, which, being but lately out of their nests, had not acquired those powers and command of wing that enable them, when more mature, to set such enemies at defiance.

Letter 44

Selborne, Nov. 30, 1780.

Dear Sir,

Every incident that occasions a renewal of our correspondence will ever be pleasing and agreeable to me.

As to the wild wood-pigeon, the *oenas*, or *vinago*, of Ray, I am much of your mind; and see no reason for making it the origin of the common house-dove: but suppose those that have advanced that opinion may have been misled by another appellation, often given to the *oenas*, which is that of stock-dove.

Unless the stock-dove in the winter varies greatly in manners from itself in summer, no species seems more unlikely to be domesticated, and to make an house-dove. We very rarely see the latter settle on trees at all, nor does it ever haunt the woods; but the former, as long as it stays with us, from November perhaps to February, lives the same wild life with the ring-dove, *palumbus torquatus*; frequents coppices and groves, supports itself chiefly by mast, and delights to roost in the tallest beeches. Could it be known in what manner stock-doves build, the doubt would be settled with me at once, provided they construct their nests on trees, like the ring-dove, as I much suspect they do.

You received, you say, last spring a stock-dove from Sussex;

and are informed that they sometimes breed in that country. But why did not your correspondent determine the place of its nidification, whether on rocks, cliffs, or trees? If he was not an adroit ornithologist I should doubt the fact, because people with us perpetually confound the stock-dove with the ring-dove.

For my own part, I readily concur with you in supposing that house-doves are derived from the small blue rock-pigeon, for many reasons. In the first place the wild stock-dove is manifestly larger than the common house-dove, against the usual rule of domestication, which generally enlarges the breed.* Again, those two remarkable black-spots on the remiges of each wing of the stock-dove, which are so characteristic of the species, would not, one should think, be totally lost by its being reclaimed; but would often break out among its descendants. But what is worth an hundred arguments is, the instance you give in Sir Roger Mostyn's house-doves in Caernarvonshire; which, though tempted by plenty of food and gentle treatment, can never be prevailed on to inhabit their cote* for any time; but, as soon as they begin to breed, betake themselves to the fastnesses of Ormshead, and deposit their young in safety amidst the inaccessible caverns, and precipices of that stupendous promontory.

> Naturam expellas furca . . . tamen usque recurret.
>
> [Nature can be driven out with a pitchfork . . . but
> she will always return.   Horace, *Epistles* 1. 10]

I have consulted a sportsman, now in his seventy-eighth year, who tells me that fifty or sixty years back, when the beechen woods were much more extensive than at present, the number of wood-pigeons was astonishing; that he has often killed near twenty in a day; and that with a long wild-fowl piece he has shot seven or eight at a time on the wing as they came wheeling over his head: he moreover adds, which I was not aware of, that often there were among them little parties of small blue doves, which he calls *rockiers*. The food of these numberless emigrants was beech-mast and some acorns; and particularly barley, which they collected in the stubbles. But of late years, since the vast increase of turnips, that vegetable has furnished a great part of their support in hard weather; and the holes they pick in these roots greatly damage the crop. From this food their flesh has contracted a rancidness which occasions them to be rejected by nicer judges of eating, who thought them before a delicate dish. They were shot not only as they were feeding in the fields, and especially in snowy weather, but also at the close of the evening, by men who lay in ambush among the woods and groves to kill them as they came in to roost.[43] These are the principal circumstances relating to this wonderful internal migration, which with us takes place towards the end of November, and ceases early in the spring. Last winter we had in Selborne high wood about an hundred of these doves; but in former times the flocks were so vast, not only with us but all the district around, that on mornings and evenings they traversed the air, like rooks, in strings, reaching for a mile together. When they thus rendez-voused here by thousands, if they happened to be suddenly roused from their roost-trees on an evening,

> Their rising all at once was like the sound
> Of thunder heard remote.
>
> [Milton, *Paradise Lost*, II]

It will by no means be foreign to the present purpose to add, that I had a relation in this neighbourhood who made it a practice, for a time, whenever he could procure the eggs of a ring-dove, to place them under a pair of doves that were sitting in his own pigeon-house; hoping thereby, if he could bring about a coalition, to enlarge his breed, and teach his own doves to beat out into the woods and to support themselves by mast: the plan was plausible, but something always inter-rupted the success; for though the birds were usually hatched, and sometimes grew to half their size, yet none ever arrived at maturity. I myself have seen these foundlings in their nest displaying a strange ferocity of nature, so as scarcely to bear to be looked at, and snapping with their bills by way of menace. In short, they always died, perhaps for want of proper sustenance: but the owner thought that by their fierce and wild demeanour they frighted their foster-mothers, and so were starved.

Virgil, as a familiar occurrence, by way of simile, describes a dove haunting the cavern of a rock in such engaging numbers, that I cannot refrain from quoting a passage: and John Dryden has rendered it so happily in our language, that without farther excuse I shall add his translation also.

> Qualis speluncâ subitò commota Columba,
> Cui domus, et dulces latebroso in pumice nidi,
> Fertur in arva volans, plausumque exterrita pennis
> Dat tecto ingentem—mox aere lapsa quieto,
> Radit iter liquidum, celeres neque commovet alas.

> As when a dove her rocky hold forsakes,
> Rous'd, in a fright her sounding wings she shakes;
> The cavern rings with clattering:—out she flies,
> And leaves her callow care, and cleaves the skies:
> At first she flutters:—but at length she springs
> To smoother flight, and shoots upon her wings.

> [Virgil, *Aeneid* 5; trans. Dryden]

LETTERS

TO THE

Honourable DAINES BARRINGTON

Snipe

Letter 1

Selborne, June 30, 1768.

DEAR SIR,

When I was in town last month* I partly engaged that I would sometime do myself the honour to write to you on the subject of natural history: and I am the more ready to fulfil my promise, because I see you are a gentleman of great candour, and one that will make allowances; especially where the writer professes to be an out-door naturalist, one that takes his observations from the subject itself, and not from the writings of others.

The following is a List of the SUMMER BIRDS *of* PASSAGE *which I have discovered in this neighbourhood, ranged somewhat in the order in which they appear:**

	RAII NOMINA	USUALLY APPEARS ABOUT
1. Wryneck	*Jynx, sive torquilla:*	The middle of March: harsh note.
2. Smallest willow-wren	*Regulus non cristatus:*	March 23: chirps till September.
3. Swallow	*Hirundo domestica:*	April 13.

4. Martin	*Hirundo rustica*:	April 13.
5. Sand-martin	*Hirundo riparia*:	Ditto.
6. Black-cap	*Atricapilla*:	Ditto: a sweet wild note.
7. Nightingale	*Luscinia*:	Beginning of April.
8. Cuckoo	*Cuculus*:	Middle of April.
9. Middle willow-wren	*Regulus non cristatus*:	Ditto: a sweet plaintive note.
10. White-throat	*Ficedulae affinis*:	Ditto: mean note; sings on till September.
11. Red-start	*Ruticilla*:	Ditto: more agreeable song.
12. Stone-curlew	*Oedicnemus*:	End of March: loud nocturnal whistle.
13. Turtle-dove	*Turtur*.	
14. Grasshopper-lark	*Alauda minima locustae voce*:	Middle April: a small sibilous note, till the end of July.
15. Swift	*Hirundo apus*:	About April 27.
16. Less reed-sparrow	*Passer arundinaceus minor*:	A sweet polyglot, but hurrying: it has the notes of many birds.
17. Land-rail	*Ortygometra*:	A loud harsh note, crex, crex.
18. Largest willow-wren	*Regulus non cristatus*:	*Cantat voce stridula locustae*: end of April on the tops of high beeches.
19. Goatsucker, or fern-owl	*Caprimulgus*:	Beginning of May: chatters by night with a singular noise.
20. Fly-catcher	*Stoparola*:	May 12. A very mute bird: this is the latest summer bird of passage.

This assemblage of curious and amusing birds belong to ten several genera of the Linnaean system; and are all of the *ordo* of *passeres* save the *jynx* and *cuculus*, which are *picae*, and the *charadrius* (*oedicnemus*) and *rallus* (*ortygometra*), which are *grallae*.

These birds, as they stand numerically,* belong to the following Linnaean genera:

1:	*Jynx*.	13:	*Columba*.
2, 6, 7, 9, 10, 11, 16, 18:	*Motacilla*.	17:	*Rallus*.
3, 4, 5, 15:	*Hirundo*.	19:	*Caprimulgus*.
8:	*Cuculus*.	14:	*Alauda*.
12:	*Charadrius*.	20:	*Muscicapa*.

Most soft-billed birds live on insects, and not on grain and seeds; and therefore at the end of summer they retire: but the following soft-billed birds, though insect-eaters, stay with us the year round:

	RAII NOMINA	
Redbreast	*Rubecula*:	These frequent houses, and haunt out-buildings in the winter: eat spiders.
Wren	*Passer troglodytes*:	
Hedge-sparrow	*Curruca*:	Haunt sinks for crumbs and other sweepings.
White-wagtail	*Motacilla alba*:	These frequent shallow rivulets near the spring heads, where they never freeze: eat the aureliae of *Phryganea*: the smallest birds that walk.
Yellow wagtail	*Motacilla flava*:	
Grey wagtail	*Motacilla cinerea*:	
Wheat-ear	*Oenanthe*:	Some of these are to be seen with us the winter through.
Whin-chat	*Oenanthe secunda*.	
Stone-chatter	*Oenanthe tertia*.	
Golden-crowned wren	*Regulus cristatus*:	This is the smallest British bird: haunts the tops of tall trees; stays the winter through.

A LIST *of the* WINTER BIRDS *of* PASSAGE *round this neighbourhood, ranged somewhat in the order in which they appear*:

		RAII NOMINA	
1.	Ring-ousel	*Merula torquata*:	This is a new migration, which I have lately discovered about Michaelmas week, and again about the fourteenth of March.
2.	Redwing	*Turdus iliacus*:	About old Michaelmas.
3.	Fieldfare	*Turdus pilaris*:	Though a percher by day, roosts on the ground.
4.	Royston-crow	*Cornix cinerea*:	Most frequently on downs.
5.	Woodcock	*Scolopax*:	Appears about old Michaelmas.
6.	Snipe	*Gallinago minor*:	Some snipes constantly breed with us.
7.	Jack-snipe	*Gallinago minima*.	
8.	Wood-pigeon	*Oenas*:	Seldom appears till late: not in such plenty as formerly.
9.	Wild-swan	*Cygnus ferus*:	On some large waters.
10.	Wild-goose	*Anser ferus*:	
11.	Wild-duck	*Anas torquata minor*:	On our lakes and streams.
12.	Pochard	*Anas fera fusca*:	
13.	Wigeon	*Penelope*:	
14.	Teal, breeds with us in Wolmer-forest	*Querquedula*:	
15.	Gross-beak	*Coccothraustes*:	These are only wanderers that appear occasionally, and are not observant of any regular migration.
16.	Cross-bill	*Loxia*:	
17.	Silk-tail	*Garrulus bohemicus*:	

These birds, as they stand numerically, belong to the following Linnaean genera:—

1, 2, 3:	*Turdus.*	9, 10, 11, 12, 13, 14:	*Anas.*
4:	*Corvus.*	15, 16:	*Loxia.*
5, 6, 7:	*Scolopax.*	17:	*Ampelis.*
8:	*Columba.*		

Birds that sing in the night are but few:

Nightingale	*Luscinia*:	'In shadiest covert hid.'
		Milton [*Paradise Lost*, III].
Woodlark	*Alauda arborea*:	Suspended in mid air.
Less reed-sparrow	*Passer arundinaceus minor*:	Among reeds and willows.

I should now proceed to such birds as continue to sing after Midsummer, but, as they are rather numerous, they would exceed the bounds of this paper: besides, as this is now the season for remarking on that subject, I am willing to repeat my observations on some birds concerning the continuation of whose song I seem at present to have some doubt.

I am, &c.

Letter 2

Selborne, Nov. 2,* 1769.

DEAR SIR

When I did myself the honour to write to you about the end of last June on the subject of natural history, I sent you a list of the summer-birds of passage which I have observed in this neighbourhood; and also a list of the winter-birds of passage: I mentioned besides those soft-billed birds that stay with us the winter through in the south of England, and those that are remarkable for singing in the night.

According to my proposal, I shall now proceed to such birds (singing birds strictly so called) as continue in full song till after Midsummer; and shall range them somewhat in the order in which they first begin to open as the spring advances.

RAII NOMINA

1. Wood-lark	*Alauda arborea*:	In January, and continues to sing through all the summer and autumn.
2. Song-thrush	*Turdus simpliciter dictus*:	In February and on to August: reassume their song in autumn.
3. Wren	*Passer troglodytes*:	All the year, hard frost excepted.
4. Redbreast	*Rubecula*:	Ditto.
5. Hedge-sparrow	*Curruca*:	Early in February to July the 10th.
6. Yellowhammer	*Emberiza flava*:	Early in February, and on through July to August the 21st.
7. Skylark	*Alauda vulgaris*:	In February, and on to October.
8. Swallow	*Hirundo domestica*:	From April to September.
9. Black-cap	*Atricapilla*:	Beginning of April to July 13th.
10. Titlark	*Alauda pratorum*:	From middle of April to July the 16th.
11. Blackbird	*Merula vulgaris*:	Sometimes in February and March, and so on to July the twenty-third: reassumes in autumn.
12. White-throat	*Ficedulae affinis*:	In April and on to July 23.
13. Goldfinch	*Carduelis*:	April, and through to September 16.
14. Greenfinch	*Chloris*:	On to July and August 2.
15. Less reed-sparrow	*Passer arundinaceus minor*:	May: on to beginning of July.
16. Common linnet	*Linaria vulgaris*:	Breeds and whistles on till August: reassumes its note when they begin to congregate in October, and again early before the flocks separate.

Birds that cease to be in full song, and are usually silent at or before Midsummer:

17. Middle willow-wren	*Regulus non cristatus*:	Middle of June: begins in April.
18. Redstart	*Ruticilla*:	Ditto: begins in May.
19. Chaffinch	*Fringilla*:	Beginning of June: sings first in February.
20. Nightingale	*Luscinia*:	Middle of June: sings first in April.

Birds that sing for a short time, and very early in the spring:

21. Missel-bird	*Turdus viscivorus*:	January the 2, 1770,* in February. Is called in Hampshire and Sussex the storm-cock, because its song is supposed to forebode windy wet weather; is the largest singing bird we have.
22. Great titmouse, or ox-eye	*Fringillago*:	In February, March, April: reassumes for a short time in September.

Birds that have somewhat of a note or song, and yet are hardly to be called singing birds:

RAII NOMINA

23. Golden-crowned wren	*Regulus cristatus*:	Its note as minute as its person: frequents the tops of high oaks and firs; the smallest British bird.
24. Marsh titmouse	*Parus palustris*:	Haunts great woods: two harsh sharp notes.
25. Small willow-wren	*Regulus non cristatus*:	Sings in March, and on to September.
26. Largest ditto	Ditto:	*Cantat voce stridula locustae*: from end of April to August.
27. Grasshopper-lark	*Alauda minima voce locustae*:	Chirps all night, from the middle of April to the end of July.
28. Martin	*Hirundo agrestis*:	All the breeding time: from May to September.
29. Bullfinch	*Pyrrhula*.	
30. Bunting	*Emberiza alba*:	From the end of January to July.

All singing birds, and those that have any pretensions to song, not only in Britain, but perhaps the world through, come under the Linnaean *ordo* of *passeres*.

The above-mentioned birds, as they stand numerically, belong to the following Linnaean genera:

1, 7, 10, 27:	*Alauda*.	8, 28:	*Hirundo*.
2, 11, 21:	*Turdus*.	13, 16, 19:	*Fringilla*.
3, 4, 5, 9, 12, 15, 17, 18, 20, 23, 25, 26:	*Motacilla*.	22, 24:	*Parus*.
6, 30:	*Emberiza*.	14, 29:	*Loxia*.

Birds that sing as they fly are but few:

RAII NOMINA

Skylark	*Alauda vulgaris*:	Rising, suspended, and falling.
Titlark	*Alauda pratorum*:	In its descent: also sitting on trees, and walking on the ground.
Woodlark	*Alauda arborea*:	Suspended: in hot summer nights all night long.
Blackbird	*Merula*:	Sometimes from bush to bush.
White-throat	*Ficedulae affinis*:	Uses when singing on the wing odd jerks and gesticulations.
Swallow	*Hirundo domestica*:	In soft sunny weather.
Wren	*Passer troglodytes*:	Sometimes from bush to bush.

Birds that breed most early in these parts:

Raven	*Corvus*:	Hatches in February and March.
Song-thrush	*Turdus*:	In March.
Blackbird	*Merula*:	In March.
Rook	*Cornix frugilega*:	Builds the beginning of March.
Woodlark	*Alauda arborea*:	Hatches in April.
Ring-dove	*Palumbus torquatus*:	Lays the beginning of April.

All birds that continue in full song till after Midsummer appear to me to breed more than once.

Most kinds of birds seem to me to be wild and shy somewhat in proportion to their bulk; I mean in this island, where they are much pursued and annoyed; but in Ascension Island, and many other desolate places, mariners have found fowls so unacquainted with an human figure, that they would stand still to be taken; as is the case with boobies,* &c. As an example of what is advanced, I remark that the golden-crested wren (the smallest British bird) will stand unconcerned till you come within three or four yards of it, while the bustard (*otis*), the largest British land fowl, does not care to admit a person within so many furlongs.

I am, &c.

Letter 3

Selborne, Jan. 15, 1770.

DEAR SIR,

It was no small matter of satisfaction to me to find that you were not displeased with my little *methodus* of birds. If there was any merit in the sketch, it must be owing to its punctuality. For many months I carried a list in my pocket of the birds that were to be remarked, and, as I rode or walked about my business, I noted each day the continuance or omission of each bird's song; so that I am as sure of the certainty of my facts as a man can be of any transaction whatsoever.

I shall now proceed to answer the several queries which you put in your two obliging letters, in the best manner that I am able. Perhaps Eastwick, and its environs, where you heard so few birds, is not a woodland country, and therefore not stocked with such songsters. If you will cast your eye on my last letter, you will find that many species continued to warble after the beginning of July.

The titlark and yellowhammer breed late, the latter very late; and therefore it is no wonder that they protract their song: for I lay it down as a maxim in ornithology, that as long as there is any incubation going on there is music. As to the redbreast and wren, it is well known to the most incurious observer that they whistle the year round, hard frost excepted; especially the latter.

It was not in my power to procure you a black-cap, or a less reed-sparrow or sedge-bird, alive. As the first is undoubtedly, and the last, as far as I can yet see, a summer bird of passage, they would require more nice and curious management in a cage than I should be able to give them: they are both distinguished songsters. The note of the former has such a wild sweetness that it always brings to my mind those lines in a song in *As You like It*:

> And tune his merry note
> Unto the *wild* bird's throat.

> (Shakespeare*)

The latter has a surprising variety of notes resembling the song of several other birds; but then it has also an hurrying manner, not at all to its advantage: it is notwithstanding a delicate polyglot.

It is new to me that titlarks in cages sing in the night; perhaps only caged birds do so. I once knew a tame redbreast in a cage that always sang as long as candles were in the room; but in their wild state no one supposes they sing in the night.

I should be almost ready to doubt the fact, that there are to be seen much fewer birds in July than in any former month, notwithstanding so many young are hatched daily. Sure I am that it is far otherwise with respect to the swallow tribe, which increases prodigiously as the summer advances: and I saw at the time mentioned, many hundreds of young wagtails on the banks of the Cherwell, which almost covered the meadows.* If the matter appears as you say in the other species, may it not be owing to the dams being engaged in incubation, while the young are concealed by the leaves?

Many times have I had the curiosity to open the stomachs of woodcocks and snipes; but nothing ever occurred that helped to explain to me what their subsistence might be: all that I could ever find was a soft mucus, among which lay many pellucid small gravels.

I am, &c.

Letter 4

Selborne, Feb. 19, 1770.

DEAR SIR,

Your observation that 'the cuckoo does not deposit its egg indiscriminately in the nest of the first bird that comes in its way, but probably looks out a nurse in some degree congenerous, with whom to intrust its young,' is perfectly new to me; and struck me so forcibly, that I naturally fell into a train of thought that led me to consider whether the fact was so, and what reason there was for it. When I came to recollect and inquire, I could not find that any cuckoo had ever been seen in these parts, except in the nest of the wagtail, the hedge-sparrow, the titlark, the white-throat, and the redbreast, all soft-billed insectivorous birds. The excellent Mr Willughby mentions the nest of the *palumbus* (ring-dove), and of the *fringilla* (chaffinch), birds that subsist on acorns and grains, and such hard food: but then he does not mention them as of his own knowledge; but says afterwards that he saw himself a wagtail feeding a cuckoo. It appears hardly possible that a soft-billed bird should subsist on the same food with the hard-billed: for the former have thin membranaceous stomachs suited to their soft food; while the latter, the granivorous tribe, have strong muscular gizzards, which, like mills, grind, by the help of small gravels and pebbles, what is swallowed. This proceeding of the cuckoo, of dropping its eggs as it were by chance, is such a monstrous outrage on maternal affection, one of the first great dictates of nature; and such a violence on instinct; that, had it only been related of a bird in the Brasils, or Peru, it would never have merited our belief.* But yet, should it farther appear that this simple bird, when divested of that natural στοργή [affection] that seems to raise the kind in general above themselves, and inspire them with extraordinary degrees of cunning and address, may be still endued with a more enlarged faculty of discerning what species are suitable and congenerous nursing-mothers for its disregarded eggs and young, and may deposit them only under *their* care, this would be adding wonder to wonder, and instancing, in a fresh manner, that the methods of Providence are not subjected to any mode or rule, but astonish us in new lights, and in various and changeable appearances.

What was said by a very ancient and sublime writer concerning the defect of natural affection in the ostrich, may be well applied to the bird we are talking of:

She is hardened against her young ones, as though they were not hers:
   Because God hath deprived her of wisdom, neither hath he imparted to her understanding.[1]

Query:* Does each female cuckoo lay but one egg in a season or does she drop several in different nests according as opportunity offers?

I am, &c.

Letter 5

Selborne, April 12, 1770.

DEAR SIR,

I heard many birds of several species sing last year after Midsummer; enough to prove that the summer solstice is not the period that puts a stop to the music of the woods. The yellowhammer no doubt persists with more steadiness than any other; but the woodlark, the wren, the redbreast, the swallow, the white-throat, the goldfinch, the common linnet, are all undoubted instances of the truth of what I advanced.

If this severe season does not interrupt the regularity of the summer migrations, the blackcap will be here in two or three days. I wish it was in my power to procure you one of those songsters; but I am no birdcatcher; and so little used to birds in a cage, that I fear if I had one it would soon die for want of skill in feeding.

Was your reed-sparrow, which you kept in a cage, the thick-billed reed-sparrow of the *Zoology*, p. 320; or was it the less reed-sparrow of Ray, the sedge-bird of Mr Pennant's last publication, p. 16?

As to the matter of long-billed birds growing fatter in moderate frosts, I have no doubt within myself what should be the reason. The thriving at those times appears to me to arise altogether from the gentle check which the cold throws upon insensible perspiration. The case is just the same with blackbirds, &c.; and farmers and warreners observe, the first, that their hogs fat more kindly at such times, and the latter that their rabbits are never in such good case as in a gentle frost. But when frosts are severe, and of long continuance, the case is soon altered; for then a want of food soon over-balances the repletion occasioned by a checked perspiration. I have observed, moreover, that some human constitutions are more inclined to plumpness in winter than in summer.

When birds come to suffer by severe frost, I find that the first that fail and die are the redwing-fieldfares, and then the song-thrushes.

You wonder, with good reason, that the hedge-sparrows, &c. can be induced at all to sit on the egg of the cuckoo without being scandalized at the vast disproportioned size of the supposititious egg; but the brute creation, I suppose, have very little idea of size, colour, or number. For the common hen, I know, when the fury of incubation is on her, will sit on a single shapeless stone, instead of a nest full of eggs that have been withdrawn: and, moreover, a hen turkey, in the same circumstances, would sit on in the empty nest till she perished with hunger.

I think the matter might easily be determined: whether a cuckoo lays one or two eggs, or more, in a season, by opening a female during the laying-time. If more than one was come down out of the ovary, and advanced to a good size, doubtless then she would that spring lay more than one.

I will endeavour to get a hen, and examine.

Your supposition that there may be some natural obstruction in singing birds while they are mute, and that when this is removed the song recommences, is new and bold: I wish you could discover some good grounds for this suspicion.*

I was glad you were pleased with my specimen of the *caprimulgus*, or fern-owl; you were, I find, acquainted with the bird before.*

When we meet I shall be glad to have some conversation with you concerning the proposal you make of my drawing up an account of the animals in this neighbourhood. Your partiality towards my small abilities persuades you, I fear, that I am able to do more than is in my power: for it is no small undertaking for a man unsupported and alone to begin a natural history from his own autopsia! Though there is endless room for observation in the field of nature, which is boundless, yet investigation (where a man endeavours to be sure of his facts) can make but slow progress; and all that one could collect in many years would go into a very narrow compass.

Some extracts from your ingenious 'Investigations of the difference between the present temperature of the air in Italy',* &c. have fallen in my way; and gave me great satisfaction: they have removed the objections that always arose in my mind whenever I came to the passages which you quote. Surely the judicious Virgil, when writing a didactic poem for the region of Italy, could never think of describing freezing rivers, unless such severity of weather pretty frequently occurred!

P.S. Swallows appear amidst snows and frost.

Letter 6

Selborne, May 21, 1770.

DEAR SIR,

The severity and turbulence of last month so interrupted the regular process of summer migration, that some of the birds do but just begin to shew themselves, and others are apparently thinner than usual; as the white-throat, the black-cap, the redstart, the fly-catcher. I well remember that after the very severe spring in the year 1739–40 summer birds of passage were very scarce. They come probably hither with a south-east wind, or when it blows between those points; but in that unfavourable year the winds blowed the whole spring and summer through from the opposite quarters. And yet amidst all these disadvantages two swallows, as I mentioned in my last, appeared this year as early as the eleventh of April amidst frost and snow; but they withdrew again for a time.*

I am not pleased to find that some people seem so little satisfied with Scopoli's new publication;[2] there is room to expect great things from the hands of that man, who is a good naturalist: and one would think that an history of the birds of so distant and southern a region as Carniola would be new and interesting. I could wish to see that work, and hope to get it sent down. Dr Scopoli is physician to the wretches that work in the quicksilver mines of that district.

When you talked of keeping a reed-sparrow, and giving it seeds, I could not help wondering; because the reed-sparrow which I mentioned to you (*passer arundinaceus minor Raii*) is a soft-billed bird; and most probably migrates hence before winter; whereas the bird you kept (*passer torquatus Raii*) abides all the year, and is a thick-billed bird. I question whether the latter be much of a songster; but in this matter I want to be better informed. The former has a variety of hurrying notes, and sings all night. Some part of the song of the former, I suspect, is attributed to the latter. We have plenty of the soft-billed sort; which Mr Pennant had entirely left out of his *British Zoology*, till I reminded him of his omission. See *British Zoology* last published, p. 16.[3]

I have somewhat to advance on the different manners in which different birds fly and walk; but as this is a subject that I have not enough considered, and is of such a nature as not to be contained in a small space, I shall say nothing further about it at present.[4]

No doubt the reason why the sex of birds in their first plumage is so difficult to be distinguished is, as you say, 'because they are not to pair and discharge their parental functions till the ensuing spring'. As colours seem to be the chief external sexual distinction in many birds, these colours do not take place till sexual attachments begin to obtain. And the case is the same in quadrupeds; among whom, in their younger days, the sexes differ but little: but, as they advance to maturity, horns and shaggy manes, beards and brawny necks, &c. &c. strongly discriminate the male from the female. We may instance still farther in our own species, where a beard and stronger features are usually characteristic of the male sex: but this sexual diversity does not take place in earlier life; for a beautiful youth shall be so like a beautiful girl that the difference shall not be discernible:

> Quem si puellarum insereres choro,
> Mire sagaces falleret hospites
> Discrimen obscurum, solutis
> Crinibus, ambiguoque vultu.

[If you put him in a group of girls, those who did not know him, no matter how observant, would not be able to notice that, disguised by his hair and his boy-girl face, he differed from the rest.   Horace, *Odes*, 2. 5]

Letter 7

Ringmer, near Lewes, Oct. 8, 1770.

DEAR SIR,

I am glad to hear that Kuckalm is to furnish you with the birds of Jamaica;* a sight of the hirundines of that hot and distant island would be a great entertainment to me.

The *Anni* of Scopoli are now in my possession; and I have read the *Annus Primus* with satisfaction: for though some parts of this work are exceptionable, and he may advance some mistaken observations; yet the ornithology of so distant a country as Carniola is very curious. Men that undertake only one district are much more likely to advance natural knowledge than those that grasp at more than they can possibly be acquainted with: every kingdom, every province, should have its own monographer.

The reason perhaps why he mentions nothing of Ray's *Ornithology* may be the extreme poverty and distance of his country, into which the works of our great naturalist may have never yet found their way. You have doubts, I know, whether this *Ornithology* is genuine, and really the work of Scopoli: as to myself, I think I discover strong tokens of authenticity; the style corresponds with that of his *Entomology*; and his characters of his Ordines and Genera are many of them new, expressive, and masterly. He has ventured to alter some of the Linnaean *genera* with sufficient shew of reason.

It might perhaps be mere accident that you saw so many swifts and no swallows at Staines; because, in my long observation of those birds I never could discover the least degree of rivalry or hostility between the species.

Ray remarks that birds of the *gallinae* order, as cocks and hens, partridges and pheasants, &c. are *pulveratrices*, such as dust themselves, using that method of cleansing their feathers, and ridding themselves of their vermin. As far as I can observe, many birds that dust themselves never wash: and I once thought that those birds that wash themselves would never dust; but here I find myself mistaken; for common house-sparrows are great *pulveratrices*, being frequently seen grovelling and wallowing in dusty roads; and yet they are great washers. Does not the skylark dust?

Query: Might not Mahomet and his followers take one method of purification from these *pulveratrices*? because I find, from travellers of credit, that if a strict mussulman is journeying in a sandy desert where no water is to be found, at stated hours he strips off his clothes, and most scrupulously rubs his body over with sand or dust.

A countryman told me he had found a young fern-owl in the nest of a small bird on the ground; and that it was fed by the little bird. I went to see this extraordinary phenomenon, and found that it was a young cuckoo hatched in the nest of a titlark:* it was become vastly too big for its nest, appearing

> . . . in tenui re
> Majores pennas nido extendisse . . .
>
> [to spread its wings too wide for the
> nest.   Horace, *Epistles*, 1. 20]

and was very fierce and pugnacious, pursuing my finger, as I teazed it, for many feet from the nest, and sparring and buffeting with its wings like a game-cock. The dupe of a dam appeared at a distance, hovering about with meat in its mouth, and expressing the greatest solicitude.

In July I saw several cuckoos skimming over a large pond; and found, after some observation, that they were feeding on the *libellulae*, or dragon-flies; some of which they caught as they settled on the weeds, and some as they were on the wing. Notwithstanding what Linnaeus says, I cannot be induced to believe that they are birds of prey.*

This district affords some birds that are hardly ever heard of at Selborne. In the first place considerable flocks of cross-beaks (*loxiae curvirostrae*) have appeared this summer in the pine-groves belonging to this house; the water-ousel is said to haunt the mouth of the Lewes river, near Newhaven; and the Cornish chough builds, I know, all along the chalky cliffs of the Sussex shore.

I was greatly pleased to see little parties of ring-ousels (my newly discovered migraters) scattered, at intervals, all along the Sussex downs from Chichester to Lewes. Let them come from whence they will, it looks very suspicious that they are cantoned along the coast in order to pass the channel where severe weather advances. They visit us again in April, as it should seem, in their return; and are not to be found in the dead of winter. It is remarkable that they are very tame, and seem to have no manner of apprehensions of danger from a person with a gun. There are bustards on the wide downs near Brighthelmstone. No doubt you are acquainted with the Sussex downs: the prospects and rides round Lewes are most lovely!

As I rode along near the coast I kept a very sharp look out in the lanes and woods, hoping I might, at this time of the year, have discovered some of the summer short-winged birds of passage crowding towards the coast in order for their departure: but it was very extraordinary that I never saw a redstart, white-throat, black-cap, uncrested wren, fly-catcher, &c. And I remember to have made the same remark in former years, as I usually come to this place annually about this time. The birds most common along the coast at present are the stone-chatters, whinchats, buntings, linnets, some few wheat-ears, titlarks, &c. Swallows and house-martins abound yet, induced to prolong their stay by this soft, still, dry season.

A land tortoise, which has been kept for thirty years in a little walled court belonging to the house where I am now visiting, retires under ground about the middle of November, and comes forth again about the middle of April. When it first appears in the spring, it discovers very little inclination towards food; but in the height of summer grows voracious: and then as the summer declines its appetite declines; so that for the last six weeks in autumn it hardly eats at all. Milky plants, such as lettuces, dandelions, sowthistles, are its favourite dish. In a neighbouring village one was kept till by tradition it was supposed to be an hundred years old. An instance of vast longevity in such a poor reptile!

Letter 8

Selborne, Dec. 20, 1770.

DEAR SIR,*

There are doubtless many home internal migrations within this kingdom that want to be better understood: witness those vast flocks of hen chaffinches that appear with us in the winter without hardly any cocks among them. Now was there a due proportion of each sex, it should seem very improbable that any one district should produce such numbers of these little birds; and much more when only one half of the species appears: therefore we may conclude that the *fringillae coelebes*, for some good purposes, have a peculiar migration of their own in which the sexes part. Nor should it seem so wonderful that the intercourse of sexes in this species of birds should be interrupted in winter; since in many animals, and particularly in bucks and does, the sexes herd separately, except at the season when commerce is necessary for the continuance of the breed. For this matter of the chaffinches see *Fauna Suecica*, p. 85, and *Systema Naturae*, p. 318. I see every winter vast flights of hen chaffinches, but none of cocks.*

Your method of accounting for the periodical motions of the British singing birds, or birds of flight, is a very probable one; since the matter of food is a great regulator of the actions and proceedings of the brute creation: there is but one that can be set in competition with it, and that is love. But I cannot quite acquiesce with you in one circumstance when you advance that, 'when they have thus feasted, they again separate into small parties of five or six, and get the best fare they can within a certain district, having no inducement to go in quest of fresh-turned earth.' Now if you mean that the business of congregating is quite at an end from the conclusion of wheat-sowing to the season of barley and oats,* it is not the case with us; for larks and chaffinches, and particularly linnets, flock and congregate as much in the very dead of winter as when the husbandman is busy with his ploughs and harrows.

Sure there can be no doubt but that woodcocks and field-fares leave us in the spring, in order to cross the seas, and to retire to some districts more suitable to the purpose of breeding. That the former pair before they retire, and that the hens are forward with egg, I myself, when I was a sportsman, have often experienced.* It cannot indeed be denied but that now and then we hear of a woodcock's nest, or young birds, discovered in some part or other of this island: but then they are always mentioned as rarities, and somewhat out of the common course of things: but as to redwings and fieldfares, no sportsman or naturalist has ever yet, that I could hear, pretended to have found the nest or young of those species in any part of these kingdoms. And I the more admire at this instance as extraordinary, since, to all appearance, the same food in summer as well as in winter might support them here which maintains their congeners, the blackbirds and thrushes, did they chuse to stay the summer through. From hence it appears that it is not food alone which determines some species of birds with regard to their stay or departure. Fieldfares and redwings disappear sooner or later according as the warm weather comes on earlier or later. For I well remember, after that dreadful winter 1739–40, that cold north-east winds continued to blow on through April and May, and that these kinds of birds (what few remained of them) did not depart as usual but were seen lingering about till the beginning of June.

The best authority that we can have for the nidification of the birds above mentioned in any district, is the testimony of faunists that have written professedly the natural history of particular countries. Now, as to the fieldfare, Linnaeus, in his *Fauna Suecica* says of it that 'maximis in arboribus nidificat [it nests in the tallest trees]': and of the redwing he says, in the same place, that 'nidificat in mediis arbusculis, sive sepibus:

ova sex caeruleo-viridia maculis nigris variis [it nests in
medium-sized bushes or hedges; it lays six eggs, blue-green in
colour with black spots].' Hence we may be assured that
fieldfares and redwings breed in Sweden. Scopoli says, in his
Annus Primus, of the woodcock, that 'nupta ad nos venit circa
aequinoctium vernale [it comes to us, already mated, about
the spring equinox]': meaning in Tirol, of which he is a native.
And afterwards he adds, 'nidificat in paludibus alpinis: ova
ponit 3 . . . 5 [it nests in the Alpine swampy woods: it lays
from three to five eggs].' It does not appear from Kramer that
woodcocks breed at all in Austria: but he says 'Avis haec
septentrionalium provinciarum astivo tempore incola est; ubi
plerumque nidificat. Appropinquante hyeme australiores
provincias petit: hinc circa plenilunium mensis Octobris plerumque Austriam transmigrat. Tunc rursus circa plenilunium
potissimum mensis Martii per Austriam matrimonio juncta ad
septentrionales provincias redit [This bird lives in northern
parts in the summer, where it breeds in great numbers. When
winter approaches, it seeks southern countries, crossing Austria in considerable numbers during the October full moon.
Then, if possible about the time of the March full moon, it
returns through Austria, already mated, to the northern
regions].' For the whole passage (which I have abridged), see
Elenchus, &c. p. 351. This seems to be a full proof of the
migration of woodcocks; though little is proved concerning the
place of breeding.*

P. S. There fell in the county of Rutland, in three weeks of
this present very wet weather, seven inches and a half of rain,
which is more than has fallen in any three weeks for these
thirty years past in that part of the world. A mean quantity in
that county for one year is twenty inches and a half.*

Letter 9

Fyfield, near Andover, Feb. 12, 1771.

DEAR SIR,

You are, I know, no great friend to migration; and the well attested accounts from various parts of the kingdom seem to justify you in your suspicions, that at least many of the swallow kind do not leave us in the winter, but lay themselves up like insects and bats, in a torpid state, and slumber away the more uncomfortable months till the return of the sun and fine weather awakens them.

But then we must not, I think, deny migration in general; because migration certainly does subsist in some places, as my brother in Andalusia has fully informed me. Of the motions of these birds he has ocular demonstration, for many weeks together, both spring and fall: during which periods, myriads of the swallow kind traverse the Straits from north to south, and from south to north, according to the season. And these vast migrations consist not only of hirundines but of bee-birds, hoopoes, *oro pendolos*, or golden thrushes, &c. &c. and also of many of our soft-billed summer birds of passage; and moreover of birds which never leave us, such as all the various sorts of hawks and kites. Old Belon,* two hundred years ago, gives a curious account of the incredible armies of hawks and kites which he saw in the spring-time traversing the Thracian Bosphorus from Asia to Europe. Besides the above mentioned, he remarks that the procession is swelled by whole troops of eagles and vultures.

Now it is no wonder that birds residing in Africa should retreat before the sun as it advances, and retire to milder regions, and especially birds of prey, whose blood being heated with hot animal food, are more impatient of a sultry climate: but then I cannot help wondering why kites and hawks, and such hardy birds as are known to defy all the severity of England, and even of Sweden and all north Europe, should want to migrate from the south of Europe, and be dissatisfied with the winters of Andalusia.

It does not appear to me that much stress may be laid on

the difficulty and hazard that birds must run in their migrations, by reason of vast oceans, cross winds, &c, because, if we
reflect, a bird may travel from England to the equator without
launching out and exposing itself to boundless seas, and that
by crossing the water at Dover, and again at Gibraltar.* And
I with the more confidence advance this obvious remark,
because my brother has always found that some of his birds,
and particularly the swallow kind, are very sparing of their
pains in crossing the Mediterranean: for when arrived at
Gibraltar, they do not

> . . . Rang'd in figure, wedge their way,
> . . . And set forth
> Their airy caravan high over seas
> Flying, and over lands with mutual wing
> Easing their flight . . .

(Milton [*Paradise Lost* VII])

but scout and hurry along in little detached parties of six or
seven in a company; and sweeping low, just over the surface of
the land and water, direct their course to the opposite continent at the narrowest passage they can find. They usually
slope across the bay to the south-west, and so pass over
opposite to Tangier, which, it seems, is the narrowest space.

In former letters we have considered whether it was probable that woodcocks in moon-shiny nights cross the German
ocean from Scandinavia. As a proof that birds of less speed
may pass that sea, considerable as it is, I shall relate the
following incident, which, though mentioned to have happened
so many years ago, was strictly matter of fact:—As some
people were shooting in the parish of Trotton, in the county of
Sussex, they killed a duck in that dreadful winter 1708–9, with
a silver collar about its neck,[5] on which were engraven the
arms of the king of Denmark. This anecdote the rector of
Trotton at that time has often told to a near relation of mine;
and, to the best of my remembrance, the collar was in the
possession of the rector.

At present I do not know any body near the sea-side that
will take the trouble to remark at what time of the moon
woodcocks first come: if I lived near the sea myself I would

soon tell you more of the matter. One thing I used to observe when I was a sportsman, that there were times in which woodcocks were so sluggish and sleepy that they would drop again when flushed just before the spaniels, nay just as the muzzle of a gun that had been fired at them: whether this strange laziness was the effect of a recent fatiguing journey I shall not presume to say.

Nightingales not only never reach Northumberland and Scotland, but also, as I have been always told, Devonshire and Cornwall. In those two last counties we cannot attribute the failure of them to the want of warmth: the defect in the west is rather a presumptive argument that these birds come over to us from the continent at the narrowest passage, and do not stroll so far westward.

Let me hear from your own observation whether skylarks do not dust. I think they do: and if they do, whether they wash also.*

The *alauda pratensis* of Ray was the poor dupe that was educating the booby of a cuckoo mentioned in my letter of October last.

Your letter came too late for me to procure a ring-ousel for Mr Tunstal during their autumnal visit; but I will endeavour to get him one when they call on us again in April. I am glad that you and that gentleman saw my Andalusian birds;* I hope they answered your expectation. Royston, or grey crows, are winter birds that come much about the same time with the woodcock: they, like the fieldfare and redwing, have no apparent reason for migration; for as they fare in the winter like their congeners, so might they in all appearance in the summer. Was not Tenant,* when a boy, mistaken? did he not find a missel-thrush's nest, and take it for the nest of a fieldfare?

The stock-dove, or wood-pigeon, *oenas Raii*, is the last winter bird of passage which appears with us; and is not seen till towards the end of November: about twenty years ago they abounded in the district of Selborne; and strings of them were seen morning and evening that reached a mile or more: but since the beechen woods have been greatly thinned they are much decreased in number. The ring-dove, *palumbus Raii*, stays with us the whole year, and breeds several times through the summer.

Before I received your letter of October last I had just remarked in my journal that the trees were unusually green. This uncommon verdure lasted on late into November; and may be accounted for from a late spring, a cool and moist summer; but more particularly from vast armies of chafers, or tree-beetles, which, in many places, reduced whole woods to a leafless naked state. These trees shot again at Midsummer, and then retained their foliage till very late in the year.

My musical friend, at whose house I am now visiting, has tried all the owls that are his near neighbours with a pitch-pipe set at concert-pitch, and finds they all hoot in B flat.* He will examine the nightingales next spring.

I am, &c. &c.

Letter 10

Selborne, Aug. 1, 1771.

DEAR SIR,

From what follows, it will appear that neither owls nor cuckoos keep to one note. A friend remarks that many (most) of his owls hoot in B flat; but that one went almost half a note below A. The pipe he tried their notes by was a common half-crown pitch-pipe,* such as masters use for tuning of harpsichords; it was the common London pitch.

A neighbour of mine, who is said to have a nice ear, remarks that the owls about this village hoot in three different keys, in G flat or F sharp, in B flat and A flat. He heard two hooting to each other, the one in A flat, and the other in B flat. *Query*: Do these different notes proceed from different species, or only from various individuals? The same person finds upon trial that the note of the cuckoo (of which we have but one species) varies in different individuals; for, about Selborne wood, he found they were mostly in D: he heard two sing together, the one in D, the other in D sharp, who made a disagreeable concert: he afterwards heard one in D sharp, and about Wolmer-forest some in C. As to nightingales, he says that their notes are so short, and their transitions so rapid, that he cannot well ascertain their key. Perhaps in a cage, and in a room, their notes may be more distinguishable. This person has tried to settle the notes of a swift, and of several other small birds, but cannot bring them to any criterion.

As I have often remarked that redwings are some of the first birds that suffer with us in severe weather, it is no wonder at all that they retreat from Scandinavian winters: and much more the *ordo* of *grallae*, who, all to a bird, forsake the northern parts of Europe at the approach of winter. 'Grallae tanquam conjuratae unanimiter in fugam se conjiciunt; ne earum unicam quidem inter nos habitantem invenire possimus; ut enim aestate in australibus degere nequeunt ob defectum lumbricorum, terramque siccam; ita nec in frigidis ob eandem causam [Waders take flight as an entire flock, as if they have conspired together. It might not be possible to find that even one has made a home with us, for just as in summer they cannot live in southern regions—because the parched soil provides no worms, so for the same reason they cannot live in cold regions]', says Ekmarck the Swede, in his ingenious little treatise called *Migrationes Avium*, which by all means you ought to read while your thoughts run on the subject of migration. See *Amoenitates Academicae*, vol. 4, p. 565.

Birds may be so circumstanced as to be obliged to migrate in one country and not in another: but the *grallae*, (which procure their food from marshes and boggy grounds) must in winter forsake the more northerly parts of Europe, or perish for want of food.

I am glad you are making inquiries from Linnaeus concerning the woodcock: it is expected of him that he should be able to account for the motions and manner of life of the animals of his own *Fauna*.

Faunists, as you observe,* are too apt to acquiesce in bare descriptions, and a few synonyms: the reason is plain; because all that may be done at home in a man's study, but the investigation of the life and conversation of animals, is a concern of much more trouble and difficulty, and is not to be attained but by the active and inquisitive, and by those that reside much in the country.

Foreign systematics are, I observe, much too vague in their specific differences; which are almost universally constituted by one or two particular marks, the rest of the description running in general terms.* But our countryman, the excellent Mr Ray, is the only describer that conveys some precise idea in every term or word, maintaining his superiority over his followers and imitators in spite of the advantage of fresh discoveries and modern information.

At this distance of years it is not in my power to recollect at what periods woodcocks used to be sluggish or alert when I was a sportsman: but, upon my mentioning this circumstance to a friend, he thinks he has observed them to be remarkably listless against snowy foul weather: if this should be the case, then the inaptitude for flying arises only from an eagerness for food; as sheep are observed to be very intent on grazing against stormy wet evenings.

Letter 11

Selborne, Feb. 8, 1772.

DEAR SIR,

When I ride about in the winter, and see such prodigious flocks of various kinds of birds, I cannot help admiring at these congregations, and wishing that it was in my power to account for those appearances almost peculiar to the season. The two great motives which regulate the proceedings of the brute creation are love and hunger; the former incites animals to perpetuate their kind, the latter induces them to preserve individuals: whether either of these should seem to be the ruling passion in the matter of congregating is to be considered. As to love, that is out of the question at a time of the year when that soft passion is not indulged: besides, during the amorous season, such a jealousy prevails between the male birds that they can hardly bear to be together in the same hedge or field. Most of the singing and elation of spirits at that time seem to me to be the effect of rivalry and emulation: and it is to this spirit of jealousy that I chiefly attribute the equal dispersion of birds in the spring over the face of the country.

Now as to the business of food: as these animals are actuated by instinct to hunt for necessary food, they should not, one would suppose, crowd together in pursuit of sustenance at a time when it is most likely to fail;* yet such associations do take place in hard weather chiefly, and thicken as the severity increases. As some kind of self-interest and self-defence is no doubt the motive for the proceeding, may it not arise from the helplessness of their state in such rigorous seasons; as men crowd together, when under great calamities, though they know not why? Perhaps approximation may dispel some degree of cold; and a crowd may make each individual appear safer from the ravages of birds of prey and other dangers.

If I admire when I see how much congenerous birds love to congregate, I am the more struck when I see incongruous ones in such strict amity. If we do not much wonder to see a flock of rooks usually attended by a train of daws, yet it is strange that the former should so frequently have a flight of starlings for their satellites. Is it because rooks have a more discerning scent than their attendants, and can lead them to spots more productive of food? Anatomists say that rooks, by reason of two large nerves which run down between the eyes into the upper mandible, have a more delicate feeling in their beaks than other round-billed birds, and can grope for their meat when out of sight. Perhaps then their associates attend them on the motive of interest, as greyhounds wait on the motions of their finders; and as lions are said to do on the yelpings of jackalls. Lapwings and starlings sometimes associate.

Letter 12

March 9, 1772.

DEAR SIR,

As a gentleman and myself were walking on the fourth of last November round the sea-banks at Newhaven, near the mouth of the Lewes river, in pursuit of natural knowledge, we were surprised to see three house-swallows gliding very swiftly by us.* That morning was rather chilly, with the wind at north-west; but the tenor of the weather for some time before had been delicate, and the noons remarkably warm. From this incident, and from repeated accounts which I meet with, I am more and more induced to believe that many of the swallow kind do not depart from this island; but lay themselves up in holes and caverns; and do, insect-like and bat-like, come forth at mild times, and then retire again to their *latebrae*. Nor make I the least doubt but that, if I lived at Newhaven, Seaford, Brighthelmstone, or any of those towns near the chalk-cliffs of the Sussex coast, by proper observations, I should see swallows stirring at periods of the winter, when the noons were soft and inviting, and the sun warm and invigorating. And I am the more of this opinion from what I have remarked during some of our late springs, that though some swallows did make their appearance about the usual time, viz. the thirteenth or four-teenth of April, yet meeting with an harsh reception, and blustering cold north-east winds, they immediately withdrew, absconding for several days, till the weather gave them better encouragement.

Letter 13

April 12, 1772.

DEAR SIR,

While I was in Sussex last autumn my residence was at the village near Lewes, from whence I had formerly the pleasure of writing to you.* On the first of November I remarked that the old tortoise, formerly mentioned, began first to dig the ground in order to the forming its *hybernaculum*, which it had fixed on just beside a great turf of hepaticas. It scrapes out the ground with its fore-feet, and throws it up over its back with its hind; but the motion of its legs is ridiculously slow, little exceeding the hour-hand of a clock; and suitable to the composure of an animal said to be a whole month in performing one feat of copulation. Nothing can be more assiduous than this creature night and day in scooping the earth, and forcing its great body into the cavity; but, as the noons of that season proved unusually warm and sunny, it was continually interrupted, and called forth by the heat in the middle of the day; and though I continued there till the thirteenth of November, yet the work remained unfinished. Harsher weather, and frosty mornings, would have quickened its operations. No part of its behaviour ever struck me more than the extreme timidity it always expresses with regard to rain; for though it has a shell that would secure it against the wheel of a loaded cart, yet does it discover as much solicitude about rain as a lady dressed in all her best attire, shuffling away on the first sprinklings, and running its head up in a corner. If attended to, it becomes an excellent weather-glass; for as sure as it walks elate, and as it were on tiptoe, feeding with great earnestness in a morning, so sure will it rain before night. It is totally a diurnal animal, and never pretends to stir after it becomes dark. The tortoise, like other reptiles, has an arbitrary stomach as well as lungs; and can refrain from eating as well as breathing for a great part of the year. When first awakened it eats nothing; nor again in the autumn before it retires: through the height of the summer it feeds voraciously, devouring all the food that comes in its way. I was much taken with its sagacity in discerning those that do it kind offices: for, as soon as the good old lady comes in sight who has waited on it for more than thirty years, it hobbles towards its benefactress with aukward alacrity; but remains inattentive to strangers. Thus not only 'the ox knoweth his owner, and the ass his master's crib,'[6] but the most abject reptile and torpid of beings distinguishes the hand that feeds it, and is touched with the feelings of gratitude.

I am, &c. &c.

P.S. In about three days after I left Sussex the tortoise retired into the ground under the hepatica.

Letter 14

Selborne, March 26, 1773.

DEAR SIR,

The more I reflect on the στοργή [natural affection] of animals, the more I am astonished at its effects. Nor is the violence of this affection more wonderful than the shortness of its duration. Thus every hen is in her turn the virago of the yard, in proportion to the helplessness of her brood; and will fly in the face of a dog or a sow in defence of those chickens, which in a few weeks she will drive before her with relentless cruelty.

This affection sublimes the passions, quickens the invention, and sharpens the sagacity of the brute creation. Thus an hen, just become a mother, is no longer that placid bird she used to be, but with feathers standing an end, wings hovering, and clocking note, she runs about like one possessed. Dams will throw themselves in the way of the greatest danger in order to avert it from their progeny. Thus a partridge will tumble along before a sportsman in order to draw away the dogs from her helpless covey. In the time of nidification the most feeble birds will assault the most rapacious. All the hirundines of a village are up in arms at the sight of an hawk, whom they will persecute till he leaves that district. A very exact observer has often remarked that a pair of ravens nesting in the rock of Gibraltar would suffer no vulture or eagle to rest near their station, but would drive them from the hill with an amazing fury: even the blue thrush at the season of breeding would dart out from the clefts of the rocks to chase away the kestril, or the sparrow-hawk. If you stand near the nest of a bird that has young, she will not be induced to betray them by an inadvertent fondness, but will wait about at a distance with meat in her mouth for an hour together.

Should I farther corroborate what I have advanced above

by some anecdotes which I probably may have mentioned before in conversation,* yet you will, I trust, pardon the repetition for the sake of the illustration.

The flycatcher of the *Zoology* (the *stoparola* of Ray), builds ever year in the vines that grow on the walls of my house. A pair of these little birds had one year inadvertently placed their nest on a naked bough, perhaps in a shady time, not being aware of the inconvenience that followed. But an hot sunny season coming on before the brood was half fledged, the reflection of the wall became insupportable, and must inevitably have destroyed the tender young, had not affection suggested an expedient, and prompted the parent-birds to hover over the nest all the hotter hours, while with wings expanded, and mouths gaping for breath, they screened off the heat from their suffering offspring.

A farther instance I once saw of notable sagacity in a willow-wren, which had built in a bank in my fields. This bird a friend and myself had observed as she sat in her nest; but were particularly careful not to disturb her, though we saw she eyed us with some degree of jealousy. Some days after as we passed that way we were desirous of remarking how this brood went on; but no nest could be found, till I happened to take up a large bundle of long green moss, as it were, carelessly thrown over the nest in order to dodge the eye of any impertinent intruder.

A still more remarkable mixture of sagacity and instinct occurred to me one day as my people were pulling off the lining of a hotbed, in order to add some fresh dung. From out of the side of this bed leaped an animal with great agility that made a most grotesque figure; nor was it without great difficulty that it could be taken; when it proved to be a large white-bellied field-mouse with three or four young clinging to her teats by their mouths and feet. It was amazing that the desultory and rapid motions of this dam should not oblige her litter to quit their hold, especially when it appeared that they were so young as to be both naked and blind!

To these instances of tender attachment, many more of which might be daily discovered by those that are studious of nature, may be opposed that rage of affection, that monstrous perversion of the στοργή, which induces some females of the brute creation to devour their young because their owners have handled them too freely, or removed them from place to place! Swine, and sometimes the more gentle race of dogs and cats, are guilty of this horrid and preposterous* murder. When I hear now and then of an abandoned mother that destroys her offspring, I am not so much amazed; since reason perverted, and the bad passions let loose, are capable of any enormity: but why the parental feelings of brutes, that usually flow in one most uniform tenor, should sometimes be so extravagantly diverted, I leave to abler philosophers than myself to determine.

I am, &c.

Letter 15

Selborne, July 8, 1773.

DEAR SIR,

Some young men went down lately to a pond on the verge of Wolmer-forest to hunt flappers, or young wild-ducks, many of which they caught, and, among the rest, some very minute yet well-fledged wild-fowls alive, which upon examination I found to be teals. I did not know till then that teals ever bred in the south of England, and was much pleased with the discovery: this I look upon as a great stroke in natural history.

We have had, ever since I can remember, a pair of white owls that constantly breed under the eaves of this church. As I have paid good attention to the manner of life of these birds during their season of breeding, which lasts the summer through, the following remarks may not perhaps be unacceptable:—About an hour before sunset (for then the mice begin to run) they sally forth in quest of prey, and hunt all round the hedges of meadows and small enclosures for them, which seem to be their only food. In this irregular country we can stand on an eminence and see them beat the fields over like a setting-dog, and often drop down in the grass or corn. I have minuted these birds with my watch for an hour together, and have found that they return to their nest, the one or the other of them, about once in five minutes; reflecting at the same time on the adroitness that every animal is possessed of as far as regards the well being of itself and offspring. But a piece of address, which they show when they return loaded, should not, I think, be passed over in silence.—As they take their prey with their claws, so they carry it in their claws to their nest: but, as the feet are necessary in their ascent under the tiles, they constantly perch first on the roof of the chancel, and shift the mouse from their claws to their bill, that the feet may be at liberty to take hold of the plate on the wall as they are rising under the eaves.

White owls seem not (but in this I am not positive) to hoot at all: all that clamorous hooting appears to me to come from the wood kinds. The white owl does indeed snore and hiss in a tremendous manner; and these menaces well answer the intention of intimidating: for I have known a whole village up in arms on such an occasion, imagining the church-yard to be full of goblins and spectres. White owls also often scream horribly as they fly along; from this screaming probably arose the common people's imaginary species of screech-owl, which they superstitiously think attends the windows of dying persons. The plumage of the remiges of the wings of every species of owl that I have yet examined is remarkably soft and pliant. Perhaps it may be necessary that the wings of these birds should not make much resistance or rushing, that they may be enabled to steal through the air unheard upon a nimble and watchful quarry.

While I am talking of owls, it may not be improper to mention what I was told by a gentleman of the county of Wilts. As they were grubbing a vast hollow pollard-ash that had been the mansion of owls for centuries, he discovered at the bottom a mass of matter that at first he could not account for. After some examination, he found that it was a congeries of the bones of mice (and perhaps of birds and bats) that had been heaping together for ages, being cast up in pellets out of the crops of many generations of inhabitants. For owls cast up the bones, fur, and feathers, of what they devour, after the manner of hawks. He believes, he told me, that there were bushels of this kind of substance.

When brown owls hoot, their throats swell as big as an hen's egg. I have known an owl of this species live a full year without any water.* Perhaps the case may be the same with all birds of prey. When owls fly they stretch out their legs behind them as a balance to their large heavy heads: for as most nocturnal birds have large eyes and ears they must have large heads to contain them. Large eyes I presume are necessary to collect every ray of light, and large concave ears to command the smallest degree of sound or noise.

I am, &c.

[Letters * * * *]

It will be proper to premise here that the sixteenth, eighteenth, twentieth, and twenty-first letters have been published already in the *Philosophical Transactions*: but as nicer observation has furnished several corrections and additions, it is hoped that the republication of them will not give offence; especially as these sheets would be very imperfect without them, and as they will be new to many readers who had no opportunity of seeing them when they made their first appearance.*

The hirundines are a most inoffensive, harmless, entertaining, social, and useful tribe of birds: they touch no fruit in our gardens; delight, all except one species, in attaching themselves to our houses; amuse us with their migrations, songs, and marvellous agility; and clear our outlets from the annoyances of gnats and other troublesome insects. Some districts in the south seas, near Guiaquil,[7] are desolated, it seems, by the infinite swarms of venomous mosquitoes, which fill the air, and render those coasts insupportable. It would be worth inquiring whether any species of hirundines is found in those regions. Whoever contemplates the myriads of insects that sport in the sunbeams of a summer evening in this country, will soon be convinced to what a degree our atmosphere would be choaked with them was it not for the friendly interposition of the swallow tribe.

Many species of birds have their peculiar lice; but the hirundines alone seem to be annoyed with dipterous insects, which infest every species, and are so large, in proportion to themselves, that they must be extremely irksome and injurious to them. These are the *hippoboscae hirundinis*, with narrow subulated wings, abounding in every nest; and are hatched by the warmth of the bird's own body during incubation, and crawl about under its feathers.*

A species of them is familiar to horsemen in the south of England under the name of forest-fly; and to some of side-fly, from its running sideways like a crab. It creeps under the tails, and about the groins, of horses, which, at their first coming out of the north, are rendered half frantic by the tickling sensation; while our own breed little regards them.

The curious Réaumur discovered the large eggs, or rather *pupae*, of these flies as big as the flies themselves, which he hatched in his own bosom. Any person that will take the trouble to examine the old nests of either species of swallows may find in them the black shining cases or skins of the *pupae* of these insects: but for other particulars, too long for this place, we refer the reader to *l'Histoire d'Insectes* of that admirable entomologist,* *Tom.* iv, pl. 11.

Letter 16

Selborne, Nov. 20, 1773.

Dear Sir,

In obedience to your injunctions I sit down to give you some account of the house-martin, or martlet; and, if my monography of this little domestic and familiar bird should happen to meet with your approbation, I may probably soon extend my inquiries to the rest of the British hirundines—the swallow, the swift, and the bank-martin.

A few house-martins begin to appear about the sixteenth of April; usually some few days later than the swallow. For some time after they appear the hirundines in general pay no attention to the business of nidification, but play and sport about, either to recruit from the fatigue of their journey, if they do migrate at all, or else that their blood may recover its true tone and texture after it has been so long benumbed by the severities of winter. About the middle of May, if the weather be fine, the martin begins to think in earnest of providing a mansion for its family. The crust or shell of this nest seems to be formed of such dirt or loam as comes most readily to hand, and is tempered and wrought together with little bits of broken straws to render it tough and tenacious. As this bird often builds against a perpendicular wall without any projecting ledge under, it requires its utmost efforts to get the first foundation firmly fixed, so that it may safely carry the super-structure. On this occasion the bird not only clings with its claws, but partly supports itself by strongly inclining its tail against the wall, making that a fulcrum; and thus steadied it works and plasters the materials into the face of the brick or stone. But then, that this work may not, while it is soft and green, pull itself down by its own weight, the provident architect has prudence and forbearance enough not to advance her work too fast; but by building only in the morning, and by dedicating the rest of the day to food and amusement, gives it sufficient time to dry and harden. Thus careful workmen when they build mud-walls (informed at first perhaps by this little bird) raise but a moderate layer at a time, and then desist; lest the work should become top-heavy, and so be ruined by its own weight. By this method in about ten or twelve days is formed an hemispheric nest with a small aperture towards the top, strong, compact, and warm; and perfectly fitted for all the purposes for which it was intended. But then nothing is more common than for the house-sparrow, as soon as the shell is finished, to seize on it as its own, to eject the owner, and to line it after its own manner.

After so much labour is bestowed in erecting a mansion, as Nature seldom works in vain, martins will breed on for several years together in the same nest, where it happens to be well sheltered and secure from the injuries of weather. The shell or crust of the nest is a sort of rustic-work full of knobs and protuberances on the outside: nor is the inside of those that I have examined smoothed with any exactness at all; but is rendered soft and warm, and fit for incubation, by a lining of small straws, grasses, and feathers; and sometimes by a bed of moss interwoven with wool. In this nest they tread, or engen-der, frequently during the time of building; and the hen lays from three to five white eggs.

At first when the young are hatched, and are in a naked and helpless condition, the parent birds, with tender assiduity, carry out what comes away from their young. Were it not for this affectionate cleanliness the nestlings would soon be burnt up and destroyed in so deep and hollow a nest, by their own caustic excrement. In the quadruped creation the same neat precaution is made use of; particularly among dogs and cats, where the dams lick away what proceeds from their young. But in birds there seems to be a particular provision, that the dung of nestlings is enveloped in a tough kind of jelly, and therefore is the easier conveyed off without soiling or daubing. Yet, as nature is cleanly in all her ways, the young perform this office for themselves in a little time by thrusting their tails out at the aperture of their nest. As the young of small birds presently arrive at their ἡλικία, or full growth, they soon become impatient of confinement, and sit all day with their heads out at the orifice, where the dams, by clinging to the nest, supply them with food from morning to night. For a time the young are fed on the wing by their parents; but the feat is done by so quick and almost imperceptible a sleight, that a person must have attended very exactly to their motions before he would be able to perceive it. As soon as the young are able to shift for themselves, the dams immediately turn their thoughts to the business of a second brood: while the first flight, shaken off and rejected by their nurses, congregate in great flocks, and are the birds that are seen clustering and hovering on sunny mornings and evenings round towers and steeples, and on the roofs of churches and houses. These congregatings usually begin to take place about the first week in August; and therefore we may conclude that by that time the first flight is pretty well over. The young of this species do not quit their abodes all together; but the more forward birds get abroad some days before the rest. These approaching the eaves of buildings, and playing about before them, make people think that several old ones attend one nest. They are often capricious in fixing on a nesting-place, beginning many edifices, and leaving them unfinished; but when once a nest is completed in a sheltered place, it serves for several seasons. Those which breed in a ready finished house get the start in hatching of those that build new by ten days or a fortnight. These industrious artificers are at their labours in the long days before four in the morning: when they fix their materials they plaster them on with their chins, moving their heads with a quick vibratory motion. They dip and wash as they fly sometimes in very hot weather, but not so frequently as swallows. It has been observed that martins usually build to a north-east or north-west aspect, that the heat of the sun may not crack and destroy their nests: but instances are also remembered where they bred for many years in vast abundance in an hot stifled inn-yard, against a wall facing to the south.

Birds in general are wise in their choice of situation: but in this neighbourhood every summer is seen a strong proof to the contrary at an house without eaves in an exposed district, where some martins build year by year in the corners of the windows. But, as the corners of these windows (which face to the south-east and south-west) are too shallow, the nests are washed down every hard rain; and yet these birds drudge on to no purpose from summer to summer, without changing their aspect or house. It is a piteous sight to see them labouring when half their nest is washed away and bringing dirt ... 'generis lapsi sarcire ruinas [to repair the ruin of their fallen state. Virgil, *Georgics* 4]'. Thus is instinct a most wonderfully unequal faculty; in some instances so much above reason, in other respects so far below it! Martins love to frequent towns, especially if there are great lakes and rivers at hand; nay they even affect the close air of London.* And I have not only seen them nesting in the Borough, but even in the Strand and Fleet-street; but then it was obvious from the dinginess of their aspect that their feathers partook of the filth of that sooty atmosphere. Martins are by far the least agile of the four species; their wings and tails are short, and therefore they are not capable of such surprising turns and quick and glancing evolutions as the swallow. Accordingly they make use of a placid easy motion in a middle region of the air, seldom mounting to any great height, and never sweeping long together over the surface of the ground or water. They do not wander far for food, but affect sheltered districts, over some lake, or under some hanging wood, or in some hollow vale, especially in windy weather. They breed the latest of all the swallow kind: in 1772 they had nestlings on to October the twenty-first, and are never without unfledged young as late as Michaelmas.

As the summer declines the congregating flocks increase in numbers daily by the constant accession of the second broods; till at last they swarm in myriads upon myriads round the villages on the Thames, darkening the face of the sky as they frequent the aits of that river, where they roost. They retire, the bulk of them I mean, in vast flocks together about the beginning of October: but have appeared of late years in a considerable flight in this neighbourhood, for one day or two, as late as November the third and sixth, after they were supposed to have been gone for more than a fortnight. They therefore withdraw with us the latest of any species. Unless these birds are very short-lived indeed, or unless they do not return to the district where they are bred, they must undergo vast devastations some how, and some where; for the birds that return yearly bear no manner of proportion to the birds that retire.

House-martins are distinguished from their congeners by having their legs covered with soft downy feathers down to their toes. They are no songsters; but twitter in a pretty inward soft manner in their nests. During the time of breeding they are often greatly molested with fleas.

I am, &c.

Letter 17

Ringmer, near Lewes, Dec. 9, 1773.

DEAR SIR,

I received your last favour just as I was setting out for this place; and am pleased to find that my monography met with your approbation. My remarks are the result of many years' observation; and are, I trust, true in the whole: though I do not pretend to say that they are perfectly void of mistake, or that a more nice observer might not make any additions, since subjects of this kind are inexhaustible.

If you think my letter worthy the notice of your respectable society, you are at liberty to lay it before them;* and they will consider it, I hope, as it was intended, as an humble attempt to promote a more minute inquiry into natural history; into the life and conversation of animals. Perhaps hereafter I may be induced to take the house-swallow under consideration; and from that proceed to the rest of the British hirundines.

Though I have now travelled the Sussex-downs upwards of thirty years, yet I still investigate that chain of majestic mountains with fresh admiration year by year; and I think I see new beauties every time I traverse it. The range, which runs from Chichester eastward as far as East-Bourn, is about sixty miles in length, and is called The South Downs, properly speaking, only round Lewes. As you pass along you command a noble view of the wild, or weald, on one hand, and the broad downs and sea on the other. Mr Ray used to visit a family[8] just at the foot of these hills, and was so ravished with the prospect from Plumpton-plain, near Lewes, that he mentions those scapes in his *Wisdom of God in the Works of the Creation* with the utmost satisfaction,* and thinks them equal to any thing he had seen in the finest parts of Europe.

For my own part, I think there is somewhat peculiarly sweet and amusing in the shapely figured aspect of chalk-hills in preference to those of stone, which are rugged, broken, abrupt, and shapeless.

Perhaps I may be singular in my opinion, and not so happy as to convey to you the same idea; but I never contemplate these mountains without thinking I perceive somewhat analogous to growth in their gentle swellings and smooth fungus-like protuberances, their fluted sides, and regular hollows and slopes, that carry at once the air of vegetative dilatation and expansion ... Or was there ever a time when these immense masses of calcarious matter were thrown into fermentation by some adventitious moisture; were raised and leavened into such shapes by some plastic power;* and so made to swell and heave their broad backs into the sky so much above the less animated clay of the wild below?

By what I can guess from the admeasurements of the hills that have been taken round my house, I should suppose that these hills surmount the wild at an average at about the rate of five hundred feet.

One thing is very remarkable as to the sheep: from the westward till you get to the river Adur all the flocks have horns, and smooth white faces, and white legs; and a hornless sheep is rarely to be seen: but as soon as you pass that river eastward, and mount Beeding-hill, all the flocks at once become hornless, or, as they call them, poll-sheep; and have moreover black faces with a white tuft of wool on their foreheads, and speckled and spotted legs: so that you would think that the flocks of Laban were pasturing on one side of the stream, and the variegated breed of his son-in-law Jacob were cantoned along on the other. And this diversity holds good respectively on each side from the valley of Bramber and Beeding to the eastward, and westward all the whole length of the downs.* If you talk with the shepherds on this subject, they tell you that the case has been so from time immemorial; and smile at your simplicity if you ask them whether the situation of these two different breeds might not be reversed? However, an intelligent friend of mine near Chichester is determined to try the experiment; and has this autumn, at the hazard of being laughed at, introduced a parcel of black-faced hornless rams among his horned western ewes. The black-faced poll-sheep have the shortest legs and the finest wool.

As I had hardly ever before travelled these downs at so late a season of the year, I was determined to keep as sharp a look-out as possible so near the southern coast, with respect to the summer short-winged birds of passage. We make great inquiries concerning the withdrawing of the swallow kind, without examining enough into the causes why this tribe is never to be seen in winter: for, *entre nous*, the disappearing of the latter is more marvellous than that of the former, and much more unaccountable. The hirundines, if they please, are certainly capable of migration; and yet no doubt are often found in a torpid state: but redstarts, nightingales, white-throats, black-caps, &c. &c. are very ill provided for long flights; have never been once found, as I ever heard of, in a torpid state, and yet can never be supposed, in such troops, from year to year to dodge and elude the eyes of the curious and inquisitive, which from day to day discern the other small birds that are known to abide our winters. But, notwithstanding all my care, I saw nothing like a summer bird of passage: and, what is more strange, not one wheat-ear, though they abound so in the autumn as to be a considerable perquisite to the shepherds that take them; and though many are seen to my knowledge all the winter through in many parts of the south of England. The most intelligent shepherds tell me that some few of these birds appear on the downs in March, and then withdraw to breed probably in warrens and stone-quarries: now and then a nest is ploughed up in a fallow on the downs under a furrow, but it is thought a rarity. At the time of wheat-harvest they begin to be taken in great numbers; are sent for sale in vast quantities to Brighthelmstone and Tunbridge; and appear at the tables of all the gentry that entertain with any degree of elegance. About Michaelmas they retire and are seen no more till March. Though these birds are, when in season, in great plenty on the south downs round Lewes, yet at East-Bourn, which is the eastern extremity of those downs, they abound much more. One thing is very remarkable—that though in the height of the season so many hundreds of dozens are taken, yet they never are seen to flock; and it is a rare thing to see more than three or four at a time: so that there must be a perpetual flitting and constant progressive succession. It does not appear that any wheat-ears are taken to the westward of Houghton-bridge, which stands on the river Arun.

I did not fail to look particularly after my new migration of ring-ousels; and to take notice whether they continued on the downs to this season of the year; as I had formerly remarked them in the month of October all the way from Chichester to Lewes wherever there were any shrubs and covert: but not one bird of this sort came within my observation. I only saw a few larks and whinchats, some rooks, and several kites and buzzards.

About Midsummer a flight of cross-bills comes to the pine-groves about this house, but never makes any long stay.

The old tortoise, that I have mentioned in a former letter, still continues in this garden; and retired under ground about the twentieth of November, and came out again for one day on the thirtieth: it lies now buried in a wet swampy border under a wall facing to the south, and is enveloped at present in mud and mire!

Here is a large rookery round this house, the inhabitants of which seem to get their livelihood very easily; for they spend the greatest part of the day on their nest-trees when the weather is mild.* These rooks retire every evening all the winter from this rookery, where they only call by the way, as they are going to roost in deep woods: at the dawn of day they always revisit their nest-trees, and are preceded a few minutes by a flight of daws, that act, as it were, as their harbingers.

I am, &c.

Letter 18

Selborne, Jan. 29, 1774.

DEAR SIR,

The house-swallow, or chimney-swallow, is undoubtedly the first comer of all the British hirundines; and appears in general on or about the thirteenth of April, as I have remarked from many years' observation. Not but now and then a straggler is seen much earlier: and, in particular, when I was a boy I observed a swallow for a whole day together on a sunny warm Shrove Tuesday; which day could not fall out later than the middle of March, and often happened early in February.*

It is worth remarking that these birds are seen first about lakes and mill-ponds; and it is also very particular, that if these early visiters happen to find frost and snow, as was the case of the two dreadful springs of 1770 and 1771, they immediately withdraw for a time. A circumstance this much more in favour of hiding than migration; since it is much more probable that a bird should retire to its hybernaculum just at hand, than return for a week or two only to warmer latitudes.

The swallow, though called the chimney-swallow, by no means builds altogether in chimneys, but often within barns

Swallow

and out-houses against the rafters; and so she did in Virgil's time:

> ... Ante
> Garrula quam tignis nidos suspendat hirundo.
>
> [Before the twittering swallow hangs its nest
> from the rafters.   Virgil, *Georgics* 4]

In Sweden she builds in barns, and is called *ladu swala*, the barn swallow. Besides, in the warmer parts of Europe there are no chimneys to houses, except they are English-built: in these countries she constructs her nest in porches, and gate-ways, and galleries, and open halls.

Here and there a bird may affect some odd, peculiar place; as we have known a swallow build down the shaft of an old well, through which chalk had been formerly drawn up for the purpose of manure: but in general with us this *hirundo* breeds in chimnies, and loves to haunt those stacks where there is a constant fire, no doubt for the sake of warmth. Not that it can subsist in the immediate shaft where there is a fire; but prefers one adjoining to that of the kitchen, and disregards the perpetual smoke of that funnel, as I have often observed with some degree of wonder.

Five or six or more feet down the chimney does this little bird begin to form her nest about the middle of May, which consists, like that of the house-martin, of a crust or shell composed of dirt or mud, mixed with short pieces of straw to render it tough and permanent; with this difference, that whereas the shell of the martin is nearly hemispheric, that of the swallow is open at the top and like half a deep dish: this nest is lined with fine grasses, and feathers which are often collected as they float in the air.

Wonderful is the address which this adroit bird shews all day long in ascending and descending with security through so narrow a pass. When hovering over the mouth of the funnel, the vibrations of her wings acting on the confined air occasion a rumbling like thunder. It is not improbable that the dam submits to this inconvenient situation so low in the shaft, in order to secure her broods from rapacious birds, and particularly from owls, which frequently fall down chimnies, perhaps in attempting to get at these nestlings.

The swallow lays from four to six white eggs, dotted with red specks; and brings out her first brood about the last week in June, or the first week in July. The progressive method by which the young are introduced into life is very amusing: first, they emerge from the shaft with difficulty enough, and often fall down into the rooms below: for a day or so they are fed on the chimney-top, and then are conducted to the dead leafless bough of some tree, where, sitting in a row, they are attended with great assiduity, and may then be called perchers. In a day or two more they become flyers, but are still unable to take their own food; therefore they play about near the place where the dams are hawking for flies; and, when a mouthful is collected, at a certain signal given, the dam and the nestling advance, rising towards each other, and meeting at an angle;

the young one all the while uttering such a little quick note of gratitude and complacency, that a person must have paid very little regard to the wonders of Nature that has not often remarked this feat.

The dam betakes herself immediately to the business of a second brood as soon as she is disengaged from her first; which at once associates with the first broods of house-martins; and with them congregates, clustering on sunny roofs, towers, and trees. This hirundo brings out her second brood towards the middle and end of August.

All the summer long is the swallow a most instructive pattern of unwearied industry and affection; for, from morning to night, while there is a family to be supported, she spends the whole day in skimming close to the ground, and exerting the most sudden turns and quick evolutions. Avenues, and long walks under hedges, and pasture-fields, and mown meadows where cattle graze, are her delight, especially if there are trees interspersed; because in such spots insects most abound. When a fly is taken a smart snap from her bill is heard, resembling the noise at the shutting of a watch-case; but the motion of the mandibles are too quick for the eye.

The swallow, probably the male bird, is the *excubitor* [sentinel] to house-martins, and other little birds, announcing the approach of birds of prey. For as soon as an hawk appears, with a shrill alarming note he calls all the swallows and martins about him; who pursue in a body, and buffet and strike their enemy till they have driven him from the village, darting down from above on his back, and rising in a perpendicular line in perfect security. This bird also will sound the alarm, and strike at cats when they climb on the roofs of houses, or otherwise approach the nests. Each species of hirundo drinks as it flies along, sipping the surface of the water; but the swallow alone, in general, washes on the wing, by dropping into a pool for many times together: in very hot weather house-martins and bank-martins dip and wash a little.

The swallow is a delicate songster, and in soft sunny weather sings both perching and flying; on trees in a kind of concert, and on chimney-tops: is also a bold flyer, ranging to distant downs and commons even in windy weather, which the other species seem much to dislike; nay, even frequenting exposed sea-port towns, and making little excursions over the salt-water. Horsemen on wide downs are often closely attended by a little party of swallows for miles together, which plays before and behind them, sweeping around, and collecting all the sculking insects that are roused by the trampling of the horses' feet: when the wind blows hard, without this expedient, they are often forced to settle to pick up their lurking prey.

This species feeds much on little *coleoptera*, as well as on gnats and flies; and often settles on dug ground, or paths, for gravels to grind and digest its food. Before they depart, for some weeks, to a bird, they forsake houses and chimnies, and roost in trees; and usually withdraw about the beginning of October; though some few stragglers may appear on at times till the first week in November.

Some few pairs haunt the new and open streets of London next the fields, but do not enter, like the house-martin, the close and crowded parts of the city.

Both male and female are distinguished from their congeners by the length and forkedness of their tails. They are undoubtedly the most nimble of all the species: and when the male pursues the female in amorous chase, they then go beyond their usual speed, and exert a rapidity almost too quick for the eye to follow.

After this circumstantial detail of the life and discerning στοργή [natural affection] of the swallow, I shall add, for your farther amusement, an anecdote or two not much in favour of her sagacity:—

A certain swallow built for two years together on the handles of a pair of garden-shears, that were stuck up against the boards in an out-house, and therefore must have her nest spoiled whenever that implement was wanted: and, what is stranger still, another bird of the same species built its nest on the wings and body of an owl that happened by accident to hang dead and dry from the rafter of a barn. This owl, with the nest on its wings, and with eggs in the nest, was brought as a curiosity worthy the most elegant private museum in Great-Britain. The owner, struck with the oddity of the sight, furnished the bringer with a large shell, or conch, desiring him to fix it just where the owl hung: the person did as he was ordered, and the following year a pair, probably the same pair, built their nest in the conch, and laid their eggs.*

The owl and the conch make a strange grotesque appearance, and are not the least curious specimens in that wonderful collection of art and nature.[9]

Thus is instinct in animals, taken the least out of its way, an undistinguishing, limited faculty; and blind to every circumstance that does not immediately respect self-preservation, or lead at once to the propagation or support of their species.

I am,

With all respect, &c. &c.

Letter 19

Selborne, Feb. 14, 1774.

DEAR SIR,

I received your favour of the eighth, and am pleased to find that you read my little history of the swallow with your usual candour: nor was I the less pleased to find that you made objections where you saw reason.

As to the quotations, it is difficult to say precisely which species of *hirundo* Virgil might intend in the lines in question, since the ancients did not attend to specific differences like modern naturalists: yet somewhat may be gathered, enough to incline me to suppose that in the two passages quoted the poet had his eye on the swallow.

In the first place the epithet *garrula* suits the swallow well, who is a great songster; and not the martin, which is rather a mute bird; and when it sings is so inward as scarce to be heard. Besides, if *tignum* in that place signifies a rafter rather than a beam, as it seems to me to do, then I think it must be the swallow that is alluded to, and not the martin; since the former does frequently build within the roof against the rafters; while the latter always, as far as I have been able to observe, builds without the roof against eaves and cornices.

As to the simile, too much stress must not be laid on it; yet the epithet *nigra* speaks plainly in favour of the swallow, whose back and wings are very black; while the rump of the martin is milk-white, its back and wings blue, and all its under part white as snow. Nor can the clumsy motions (comparatively clumsy) of the martin well represent the sudden and artful evolutions and quick turns which Juturna gave to her brother's chariot, so as to elude the eager pursuit of the enraged Aeneas. The verb *sonat* also seems to imply a bird that is somewhat loquacious.[10]

We have had a very wet autumn and winter, so as to raise the springs to a pitch beyond any thing since 1764; which was a remarkable year for floods and high waters. The land-springs, which we call *lavants*, break out much on the downs of Sussex, Hampshire, and Wiltshire. The country people say when the lavants rise corn will always be dear; meaning that when the earth is so glutted with water as to send forth springs on the downs and uplands, that the corn-vales must be drowned: and so it has proved for these ten or eleven years past. For land-springs have never obtained more since the memory of man than during that period, nor has there been known a greater scarcity of all sorts of grain considering the great improvements of modern husbandry. Such a run of wet seasons a century or two ago would, I am persuaded, have occasioned a famine. Therefore, pamphlets and newspaper letters, that talk of combinations, tend to inflame and mislead; since we must not expect plenty till Providence sends us more favourable seasons.*

The wheat of last year, all round this district, and in the county of Rutland, and elsewhere, yields remarkably bad: and our wheat on the ground, by the continual late sudden vicissitudes from fierce frost to pouring rains, looks poorly; and the turnips rot very fast.

I am, &c.

Letter 20

Selborne, Feb. 26, 1774.

DEAR SIR,

The sand-martin, or bank-martin, is by much the least of any of the British hirundines; and, as far as we have ever seen, the smallest known hirundo: though Brisson asserts that there is one much smaller, and that is the *hirundo esculenta*.*

But it is much to be regretted that it is scarce possible for any observer to be so full and exact as he could wish in reciting the circumstances attending the life and conversation of this little bird, since it is *fera naturâ*,* at least in this part of the kingdom, disclaiming all domestic attachments, and haunting wild heaths and commons where there are large lakes: while the other species, especially the swallow and house-martin, are remarkably gentle and domesticated, and never seem to think themselves safe but under the protection of man.

Here are in this parish, in the sand-pits and banks of the lakes of Wolmer-forest, several colonies of these birds; and yet they are never seen in the village; nor do they at all frequent the cottages that are scattered about in that wild district. The only instance I ever remember where this species haunts any building is at the town of Bishop's Waltham, in this county, where many sand-martins nestle and breed in the scaffold-holes of the back-wall of William of Wykeham's stables: but then this wall stands in a very sequestered and retired enclosure, and faces upon a large and beautiful lake. And indeed this species seems so to delight in large waters, that no instance occurs of their abounding, but near vast pools or rivers: and in particular it has been remarked that they swarm in the banks of the Thames in some places below London-bridge.

It is curious to observe with what different degrees of architectonic skill Providence has endowed birds of the same genus, and so nearly correspondent in their general mode of life! For while the swallow and the house-martin discover the greatest address in raising and securely fixing crusts or shells of loam as cunabula for their young, the bank-martin tere-brates a round and regular hole in the sand or earth, which is

Sand-martin

serpentine, horizontal, and about two feet deep. At the inner end of this burrow does this bird deposit, in a good degree of safety, her rude nest, consisting of fine grasses and feathers, usually goose feathers, very inartificially laid together.

Perseverance will accomplish any thing: though at first one would be disinclined to believe that this weak bird, with her soft and tender bill and claws, should ever be able to bore the stubborn sand-bank without entirely disabling herself: yet with these feeble instruments have I seen a pair of them make great dispatch: and could remark how much they had scooped that day by the fresh sand which ran down the bank, and was of a different colour from that which lay loose and bleached in the sun.

In what space of time these little artists are able to mine and finish these cavities I have never been able to discover, for reasons given above; but it would be a matter worthy of observation, where it falls in the way of any naturalist to make his remarks. This I have often taken notice of, that several holes of different depths are left unfinished at the end of summer. To imagine that these beginnings were intentionally made in order to be in the greater forwardness for next spring, is allowing perhaps too much foresight, and *rerum prudentia* [circumspection] to a simple bird. May not the cause of these *latebrae* [hiding-places] being left unfinished arise from their meeting in those places with strata too harsh, hard, and solid, for their purpose, which they relinquish, and go to a fresh spot that works more freely? Or may they not in other places fall in with a soil as much too loose and mouldering, liable to flounder, and threatening to overwhelm them and their labours?

One thing is remarkable—that, after some years, the old holes are forsaken and new ones bored; perhaps because the old habitations grow foul and fetid from long use, or because they may so abound with fleas as to become untenantable. This species of swallow moreover is strangely annoyed with fleas: and we have seen fleas, bed-fleas (*pulex irritans*),* swarming at the mouth of these holes, like bees on the stools of their hives.

The following circumstance should by no means be omitted—that these birds do not make use of their caverns by way of *hybernacula*, as might be expected; since banks so perforated have been dug out with care in the winter, when nothing was found but empty nests.

The sand-martin arrives much about the same time with the swallow, and lays, as she does, from four to six white eggs. But as this species is *cryptogame*, carrying on the business of nidification, incubation, and the support of its young in the dark, it would not be so easy to ascertain the time of breeding, were it not for the coming forth of the broods, which appear much about the time, or rather somewhat earlier than those of the swallow. The nestlings are supported in common like those of their congeners, with gnats and other small insects; and sometimes they are fed with *libellulae* (dragon-flies) almost as long as themselves. In the last week in June we have seen a row of these sitting on a rail near a great pool as perchers; and so young and helpless, as easily to be taken by hand: but whether the dams ever feed them on the wing, as swallows and house-martins do, we have never yet been able to determine; nor do we know whether they pursue and attack birds of prey.

When they happen to breed near hedges and enclosures, they are dispossessed of their breeding holes by the house-sparrow, which is on the same account a fell adversary to house-martins.

These hirundines are no songsters, but rather mute, making only a little harsh noise when a person approaches their nests. They seem not to be of a sociable turn, never with us congregating with their congeners in the autumn. Undoubtedly they breed a second time, like the house-martin and swallow; and withdraw about Michaelmas.

Though in some particular districts they may happen to abound, yet on the whole, in the south of England at least, is this much the rarest species. For there are few towns or large villages but what abound with house-martins; few churches, towers, or steeples, but what are haunted by some swifts; scarce a hamlet or single cottage-chimney that has not its swallow; while the bank-martins, scattered here and there, live a sequestered life among some abrupt sand-hills, and in the banks of some few rivers.

These birds have a peculiar manner of flying; flitting about with odd jerks, and vacillations, not unlike the motions of a butterfly. Doubtless the flight of all hirundines is influenced by, and adapted to, the peculiar sort of insects which furnish their food. Hence it would be worth inquiry to examine what particular genus of insects affords the principal food of each respective species of swallow.

Notwithstanding what has been advanced above, some few sand-martins, I see, haunt the skirts of London, frequenting the dirty pools in Saint George's-Fields, and about White-Chapel. The question is where these build, since there are no banks or bold shores in that neighbourhood:* perhaps they nestle in the scaffold holes of some old or new deserted building. They dip and wash as they fly sometimes, like the house-martin and swallow.

Sand-martins differ from their congeners in the diminutive-
ness of their size, and in their colour, which is what is usually
called a mouse-colour. Near Valencia, in Spain, they are
taken, says Willughby, and sold in the markets for the table;
and are called by the country people, probably from their
desultory jerking manner of flight, *Papilion de Montagna*.

Letter 21

Selborne, Sept. 28, 1774.

DEAR SIR,

As the swift or black-martin is the largest of the British
hirundines, so is it undoubtedly the latest comer. For I
remember but one instance of its appearing before the last
week in April: and in some of our late frosty, harsh springs, it
has not been seen till the beginning of May. This species
usually arrives in pairs.

The swift, like the sand-martin, is very defective in architecture, making no crust, or shell, for its nest; but forming it of
dry grasses and feathers, very rudely and inartificially put
together. With all my attention to these birds, I have never
been able once to discover one in the act of collecting or
carrying in materials: so that I have suspected (since their
nests are exactly the same) that they sometimes usurp upon
the house-sparrows, and expel them, as sparrows do the house
and sand-martin; well remembering that I have seen them
squabbling together at the entrance of their holes; and the
sparrows up in arms, and much-disconcerted at these intrud-
ers. And yet I am assured, by a nice observer in such matters,
that they do collect feathers for their nests in Andalusia; and
that he has shot them with such materials in their mouths.

Swifts, like sand-martins, carry on the business of nidification quite in the dark, in crannies of castles, and towers, and
steeples, and upon the tops of the walls of churches under the
roof; and therefore cannot be so narrowly watched as those
species that build more openly: but, from what I could ever
observe, they begin nesting about the middle of May; and I

have remarked, from eggs taken, that they have sat hard by the ninth of June. In general they haunt tall buildings, churches, and steeples, and breed only in such: yet in this village some pairs frequent the lowest and meanest cottages, and educate their young under those thatched roofs. We remember but one instance where they breed out of buildings; and that is in the sides of a deep chalkpit near the town of Odiham, in this county, where we have seen many pairs entering the crevices, and skimming and squeaking round the precipices.

As I have regarded these amusive birds with no small attention, if I should advance something new and peculiar with respect to them, and different from all other birds, I might perhaps be credited; especially as my assertion is the result of many years exact observation. The fact that I would advance is, that swifts tread, or copulate, on the wing: and I would wish any nice observer, that is startled at this supposition, to use his own eyes, and I think he will soon be convinced. In another class of animals, viz. the insect, nothing is so common as to see the different species of many genera in conjunction as they fly. The swift is almost continually on the wing; and as it never settles on the ground, on trees, or roofs, would seldom find opportunity for amorous rites, was it not enabled to indulge them in the air. If any person would watch these birds of a fine morning in May, as they are sailing round at a great height from the ground, he would see, every now and then, one drop on the back of another, and both of them sink down together for many fathoms with a loud piercing shriek. This I take to be the juncture when the business of generation is carrying on.

As the swift eats, drinks, collects materials for its nest, and, as it seems, propagates on the wing; it appears to live more in the air than any other bird, and to perform all functions there save those of sleeping and incubation.

This hirundo differs widely from its congeners in laying invariably but two eggs at a time, which are milk-white, long, and peaked at the small end; whereas the other species lay at each brood from four to six. It is a most alert bird, rising very early, and retiring to roost very late; and is on the wing in the height of summer at least sixteen hours. In the longest days it does not withdraw to rest until a quarter before nine in the evening, being the latest of all day birds. Just before they retire whole groups of them assemble high in the air, and squeak, and shoot about with wonderful rapidity. But this bird is never so much alive as in sultry thundry weather, when it expresses great alacrity, and calls forth all its powers. In hot mornings several, getting together into little parties, dash round the steeples and churches, squeaking as they go in a very clamorous manner: these, by nice observers, are supposed to be males serenading their sitting hens; and not without reason, since they seldom squeak till they come close to the walls or eaves, and since those within utter at the same time a little inward note of complacency.

When the hen has sat hard all day, she rushes forth just as it is almost dark, and stretches and relieves her weary limbs, and snatches a scanty meal for a few minutes, and then returns to her duty of incubation. Swifts, when wantonly and cruelly shot while they have young, discover a little lump of insects in their mouths, which they pouch and hold under their tongue. In general they feed in a much higher district than the other species; a proof that gnats and other insects do also abound to a considerable height in the air: they also range to vast distances; since loco-motion is no labour to them, who are endowed with such wonderful powers of wing. Their powers seem to be in proportion to their levers; and their wings are longer in proportion than those of almost any other bird. When they mute, or ease themselves in flight, they raise their wings, and make them meet over their backs.

At some certain times in the summer I had remarked that swifts were hawking very low for hours together over pools and streams; and could not help inquiring into the object of their pursuit that induced them to descend so much below their usual range. After some trouble, I found that they were taking *phryganeae*, *ephemerae*, and *libellulae* (cadew-flies, may-flies, and dragon flies) that were just emerged out of their aurelia state. I then no longer wondered that they should be so willing to stoop for a prey that afforded them such plentiful and succulent nourishment.

They bring out their young about the middle or latter end of July: but as these never become perchers, nor, that ever I could discern, are fed on the wing by their dams,* the coming forth of the young is not so notorious as in the other species.

On the thirtieth of last June I untiled the eaves of an house where many pairs build,* and found in each nest only two squab, naked *pulli*: on the eighth of July I repeated the same inquiry, and found they had made very little progress towards a fledged state, but were still naked and helpless. From whence we may conclude that birds whose way of life keeps them perpetually on the wing would not be able to quit their nest till the end of the month. Swallows and martins, that have numerous families, are continually feeding them every two or three minutes; while swifts, that have but two young to maintain, are much at their leisure, and do not attend on their nests for hours together.*

Sometimes they pursue and strike at hawks that come in their way; but not with that vehemence and fury that swallows express on the same occasion. They are out all day long in wet days, feeding about, and disregarding still rain: from whence two things may be gathered; first, that many insects abide high in the air, even in rain; and next, that the feathers of these birds must be well preened to resist so much wet. Windy, and particularly windy weather with heavy showers, they dislike; and on such days withdraw, and are scarce ever seen.

There is a circumstance respecting the colour of swifts, which seems not to be unworthy our attention. When they arrive in the spring they are all over of a glossy, dark soot-colour, except their chins, which are white; but, by being all day long in the sun and air, they become quite weather-beaten and bleached before they depart, and yet they return glossy again in the spring. Now, if they pursue the sun into lower latitudes, as some suppose, in order to enjoy a perpetual summer, why do they not return bleached? Do they not rather perhaps retire to rest for a season, and at that juncture moult and change their feathers, since all other birds are known to moult soon after the season of breeding?*

Swifts are very anomalous in many particulars, dissenting from all their congeners not only in the number of their young,

E.H.N

Swift

but in breeding but once in a summer; whereas all the other British hirundines breed invariably twice. It is past all doubt that swifts can breed but once, since they withdraw in a short time after the flight of their young, and some time before their congeners bring out their second broods. We may here remark that, as swifts breed but once in a summer, and only two at a time, and the other hirundines twice, the latter, who lay from four to six eggs, increase at an average five times as fast as the former.

But in nothing are swifts more singular than in their early retreat. They retire, as to the main body of them, by the tenth of August, and sometimes a few days sooner: and every straggler invariably withdraws by the twentieth, while their congeners, all of them, stay till the beginning of October; many of them all through that month, and some occasionally to the beginning of November. This early retreat is mysterious and wonderful, since that time is often the sweetest season in the year. But, what is more extraordinary, they begin to retire still earlier in the more southerly parts of Andalusia, where they can be no ways influenced by any defect of heat; or, as one might suppose, defect of food. Are they regulated in their motions with us by a failure of food, or by a propensity to moulting, or by a disposition to rest after so rapid a life, or by what? This is one of those incidents in natural history that not only baffles our searches, but almost eludes our guesses!

These hirundines never perch on trees or roofs, and so never congregate with their congeners. They are fearless while haunting their nesting places, and are not to be scared with a gun; and are often beaten down with poles and cudgels as they stoop to go under the eaves. Swifts are much infested with those pests to the genus called *hippoboscae hirundinis*; and often wriggle and scratch themselves, in their flight, to get rid of that clinging annoyance.

Swifts are no songsters, and have only one harsh screaming note; yet there are ears to which it is not displeasing, from an agreeable association of ideas, since that note never occurs but in the most lovely summer weather.

They never settle on the ground but through accident; and when down can hardly rise, on account of the shortness of their legs and the length of their wings: neither can they walk, but only crawl; but they have a strong grasp with their feet, by which they cling to walls. Their bodies being flat they can enter a very narrow crevice; and where they cannot pass on their bellies they will turn up edgewise.

The particular formation of the foot discriminates the swift from all the British hirundines; and indeed from all other known birds, the *hirundo melba*, or great white-bellied swift of Gibraltar, excepted; for it is so disposed as to carry 'omnes quatuor digitos anticos', all its four toes forward;* besides the least toe, which should be the back-toe, consists of one bone alone, and the other three only of two apiece: a construction most rare and peculiar, but nicely adapted to the purposes in which their feet are employed. This, and some peculiarities attending the nostrils and under mandible, have induced a discerning[11] naturalist to suppose that this species might constitute a *genus per se* [class in itself].

In London a party of swifts frequents the Tower, playing and feeding over the river just below the bridge: others haunt some of the churches of the Borough next the fields; but do not venture, like the house-martin, into the close crowded part of the town.

The Swedes have bestowed a very pertinent name on this swallow, calling it *ring swala* from the perpetual rings or circles that it takes round the scene of its nidification.

Swifts feed on *coleoptera*, or small beetles with hard cases over their wings, as well as on the softer insects; but it does not appear how they can procure gravel to grind their food, as swallows do, since they never settle on the ground. Young ones, over-run with *hippoboscae*, are sometimes found, under their nests, fallen to the ground; the number of vermin rendering their abode insupportable any longer. They frequent in this village several abject cottages; yet a succession still haunts the same unlikely roofs: a good proof this that the same birds return to the same spots.* As they must stoop very low to get up under these humble eaves, cats lie in wait, and sometimes catch them on the wing.

On the fifth of July, 1775, I again untiled part of a roof over the nest of a swift. The dam sat in the nest; but so strongly was she affected by natural στοργή [affection] for her brood, which she supposed to be in danger, that, regardless of her own safety, she would not stir, but lay sullenly by them, permitting herself to be taken in hand. The squab young we brought down and placed on the grass-plot, where they tumbled about, and were as helpless as a new-born child. While we contemplated their naked bodies, their unwieldly, disproportioned abdomina, and their heads, too heavy for their necks to support, we could not but wonder when we reflected that these shiftless beings in a little more than a fortnight would be able to dash through the air almost with the inconceivable swiftness of a meteor;* and perhaps, in their emigration, must traverse vast continents and oceans as distant as the equator. So soon does Nature advance small birds to their ἡλικία, or state of perfection; while the progressive growth of men and large quadrupeds is slow and tedious!

I am, &c.

Letter 22

Selborne, Sept. 13, 1774.*

DEAR SIR,

By means of a straight cottage-chimney I had an opportunity this summer of remarking, at my leisure, how swallows ascend and descend through the shaft: but my pleasure, in contemplating the address with which this feat was performed to a considerable depth in the chimney, was somewhat interrupted by apprehensions lest my eyes might undergo the same fate with those of Tobit.[12]

Perhaps it may be some amusement to you to hear at what times the different species of hirundines arrived this spring in three very distant counties of this kingdom. With us the swallow was seen first on April the 4th, the swift on April the 24th, the bank-martin on April the 12th, and the house-martin not till April the 30th. At South Zele, Devonshire, swallows did not arrive till April the 25th; swifts, in plenty, on May the 1st; and house-martins not till the middle of May. At Blackburn, in Lancashire, swifts were seen April the 28th, swallows April the 29th, house-martins May the 1st. Do these different dates, in such distant districts, prove any thing for or against migration?*

A farmer, near Weyhill, fallows his land with two teams of asses; one of which works till noon, and the other in the afternoon. When these animals have done their work, they are penned all night, like sheep, on the fallow. In the winter they are confined and foddered in a yard, and make plenty of dung.*

Linnaeus says that hawks 'paciscuntur inducias cum avibus, quamdiu cuculus cuculat [agree an armistice with other birds for the period that the cuckoo calls]': but it appears to me that, during that period, many little birds are taken and destroyed by birds of prey, as may be seen by their feathers left in lanes and under hedges.

The missel-thrush is, while breeding, fierce and pugnacious, driving such birds as approach its nest, with great fury, to a distance. The Welch call it *pen y llwyn*, the head or master of the coppice. He suffers no magpie, jay, or blackbird, to enter the garden where he haunts; and is, for the time, a good guard to the new-sown legumens.* In general he is very successful in the defence of his family: but once I observed in my garden, that several magpies came determined to storm the nest of a missel-thrush: the dams defended their mansion with great vigour, and fought resolutely *pro aris et focis* [for hearth and home. Cicero]; but numbers at last prevailed, they tore the nest to pieces, and swallowed the young alive.

In the season of nidification the wildest birds are comparatively tame. Thus the ring-dove breeds in my fields, though they are continually frequented; and the missel-thrush, though most shy and wild in the autumn and winter, builds in my garden close to a walk where people are passing all day long.

Wall-fruit abounds with me this year; but my grapes, that used to be forward and good, are at present backward beyond all precedent: and this is not the worst of the story; for the same ungenial weather, the same black cold solstice, has injured the more necessary fruits of the earth, and discoloured and blighted our wheat. The crop of hops promises to be very large.

Frequent returns of deafness incommode me sadly, and half disqualify me for a naturalist; for, when those fits are upon me, I lose all the pleasing notices and little intimations arising from rural sounds; and May is to me as silent and mute with respect to the notes of birds, &c. as August. My eyesight is, thank God, quick and good; but with respect to the other senses, I am, at times, disabled:

> And Wisdom at one entrance quite shut out.
>
> [Milton, *Paradise Lost*, iii].

Letter 23

Selborne, June 8, 1775.

DEAR SIR,

On September the 21st, 1741, being then on a visit, and intent on field-diversions, I rose before daybreak: when I came into the enclosures, I found the stubbles and clover-grounds matted all over with a thick coat of cobweb, in the meshes of which a copious and heavy dew hung so plentifully that the whole face of the country seemed, as it were, covered with two or three setting-nets drawn one over another. When the dogs attempted to hunt, their eyes were so blinded and hoodwinked that they could not proceed, but were obliged to lie down and scrape the incumbrances from their faces with their fore-feet, so that, finding my sport interrupted, I returned home musing in my mind on the oddness of the occurrence.

As the morning advanced the sun became bright and warm, and the day turned out one of those most lovely ones which no season but the autumn produces; cloudless, calm, serene, and worthy of the South of France itself.

About nine an appearance very unusual began to demand our attention, a shower of cobwebs falling from very elevated regions, and continuing, without any interruption, till the close of the day. These webs were not single filmy threads, floating in the air in all directions, but perfect flakes or rags; some near an inch broad, and five or six long, which fell with a degree of velocity that shewed they were considerably heavier than the atmosphere.

On every side as the observer turned his eyes might he behold a continual succession of fresh flakes falling into his sight, and twinkling like stars as they turned their sides towards the sun.

How far this wonderful shower extended would be difficult to say; but we know that it reached Bradley, Selborne, and Alresford, three places which lie in a sort of triangle, the shortest of whose sides is about eight miles in extent.

At the second of those places there was a gentleman (for whose veracity and intelligent turn we have the greatest veneration)* who observed it the moment he got abroad; but concluded that, as soon as he came upon the hill above his house, where he took his morning rides, he should be higher than this meteor, which he imagined might have been blown, like thistledown, from the common above: but, to his great astonishment, when he rode to the most elevated part of the down, 300 feet above his fields, he found the webs in appearance still as much above him as before; still descending into sight in a constant succession, and twinkling in the sun, so as to draw the attention of the most incurious.

Neither before nor after was any such fall observed; but on this day the flakes hung in the trees and hedges so thick, that a diligent person sent out might have gathered baskets full.

The remark that I shall make on these cobweb-like appearances, called gossamer, is, that, strange and superstitious as the notions about them were formerly, nobody in these days doubts but that they are the real production of small spiders, which swarm in the fields in fine weather in autumn, and have a power of shooting out webs from their tails so as to render themselves buoyant, and lighter than air. But why these apterous insects should *that day* take such a wonderful aërial excursion,* and why their webs should at once become so gross and material as to be considerably more weighty than air, and to descend with precipitation, is a matter beyond my skill. If I might be allowed to hazard a supposition, I should imagine that those filmy threads, when first shot, might be entangled in the rising dew, and so drawn up, spiders and all, by a brisk evaporation into the regions where clouds are formed: and if the spiders have a power of coiling and thickening their webs in the air, as Dr Lister says they have (see his Letters to Mr Ray) then, when they become heavier than the air, they must fall.

Every day in fine weather, in autumn chiefly, do I see those spiders shooting out their webs and mounting aloft: they will go off from your finger if you will take them into your hand. Last summer one alighted on my book as I was reading in the parlour; and, running to the top of the page, and shooting out a web, took its departure from thence. But what I most wondered at was, that it went off with considerable velocity in a place where no air was stirring; and I am sure that I did not assist it with my breath. So that these little crawlers seem to have, while mounting, some loco-motive power without the use of wings, and to move in the air faster than the air itself.

Letter 24

Selborne, Aug. 15, 1775.

DEAR SIR,

There is a wonderful spirit of sociality in the brute creation, independent of sexual attachment: the congregation of gregarious birds in the winter is a remarkable instance.*

Many horses, though quiet with company, will not stay one minute in a field by themselves: the strongest fences cannot restrain them. My neighbour's horse will not only not stay by himself abroad, but he will not bear to be left alone in a strange stable without discovering the utmost impatience, and endeavouring to break the rack and manger with his fore feet. He has been known to leap out at a stable-window, through which dung was thrown, after company; and yet in other respects is remarkably quiet. Oxen and cows will not fatten by themselves; but will neglect the finest pasture that is not recommended by society. It would be needless to instance in sheep, which constantly flock together.

But this propensity seems not to be confined to animals of the same species; for we know a doe, still alive, that was brought up from a little fawn with a dairy of cows; with them it goes a-field, and with them it returns to the yard. The dogs of the house take no notice of this deer, being used to her; but if strange dogs come by, a chase ensues; while the master smiles to see his favourite securely leading her pursuers over hedge, or gate, or stile, till she returns to the cows, who, with fierce lowings and menacing horns, drive the assailants quite out of the pasture.

Even great disparity of kind and size does not always prevent social advances and mutual fellowship. For a very intelligent and observant person has assured me that, in the former part of his life, keeping but one horse, he happened also on a time to have but one solitary hen. These two incongruous animals spent much of their time together in a lonely orchard, where they saw no creature but each other. By degrees an apparent regard began to take place between these two sequestered individuals. The fowl would approach the quadruped with notes of complacency, rubbing herself gently against his legs: while the horse would look down with satisfaction, and move with the greatest caution and circumspection, lest he should trample on his diminutive companion. Thus, by mutual good offices, each seemed to console the vacant hours of the other: so that Milton, when he puts the following sentiment in the mouth of Adam, seems to be somewhat mistaken:

> Much less can bird with beast, or fish with fowl,
> So well converse, nor with the ox the ape.
>
> [*Paradise Lost*, VIII].

I am, &c.

Letter 25

Selborne, Oct. 2, 1775.

DEAR SIR,

We have two gangs or hordes of gypsies which infest the south and west of England, and come round in their circuit two or three times in the year. One of these tribes calls itself by the noble name of Stanley, of which I have nothing particular to say; but the other is distinguished by an appellative somewhat remarkable—As far as their harsh gibberish can be understood, they seem to say that the name of their clan is Curleople: now the termination of this word is apparently Grecian: and as Mezeray and the gravest historians all agree that these vagrants did certainly migrate from Egypt and the East, two or three centuries ago, and so spread by degrees over Europe, may not this family-name, a little corrupted, be the very name they brought with them from the

Levant? It would be matter of some curiosity, could one meet with an intelligent person among them, to inquire whether, in their jargon, they still retain any Greek words: the Greek radicals will appear in hand, foot, head, water, earth, &c. It is possible that amidst their cant and corrupted dialect many mutilated remains of their native language might still be discovered.*

With regard to those peculiar people, the gypsies, one thing is very remarkable, and especially as they came from warmer climates; and that is, that while other beggars lodge in barns, stables, and cow-houses, these sturdy savages seem to pride themselves in braving the severities of winter, and in living *sub dio* [in the open air] the whole year round. Last September was as wet a month as ever was known; and yet during those deluges did a young gypsy-girl lie-in in the midst of one of our hop-gardens, on the cold ground, with nothing over her but a piece of blanket extended on a few hazel-rods bent hoop fashion, and stuck into the earth at each end, in circumstances too trying for a cow in the same condition: yet within this garden there was a large hop-kiln, into the chambers of which she might have retired, had she thought shelter an object worthy her attention.

Europe itself, it seems, cannot set bounds to the rovings of these vagabonds; for Mr Bell, in his return from Peking, met a gang of these people on the confines of Tartary, who were endeavouring to penetrate those deserts and try their fortune in China.[13]

Gypsies are called in French, *Bohemiens*; in Italian and modern Greek, *Zingani*.

I am, &c.

Letter 26

Selborne, Nov. 1, 1775.

DEAR SIR,

> Hic . . . taedae pingues, hic plurimus ignis
> Semper, et assidua postes fuligine nigri
>
> [Here, with the pine-wood torches full of
>   burning fat and the fire always blazing
>   high, the door-posts are black from the
>   unremitting soot.   Virgil, *Eclogues* 7]

I shall make no apology for troubling you with the detail of a
very simple piece of domestic economy, being satisfied that
you think nothing beneath your attention that tends to utility:
the matter alluded to is the use of rushes instead of candles,
which I am well aware prevails in many districts besides this;
but as I know there are countries also where it does not obtain,
and as I have considered the subject with some degree of
exactness, I shall proceed in my humble story, and leave you
to judge of the expediency.

The proper species of rush for this purpose seems to be the
juncus conglomeratus, or common soft rush,* which is to be found
in most moist pastures, by the sides of streams, and under
hedges. These rushes are in best condition in the height of
summer; but may be gathered, so as to serve the purpose well,
quite on to autumn. It would be needless to add that the
largest and longest are best. Decayed* labourers, women, and
children, make it their business to procure and prepare them.
As soon as they are cut they must be flung into water, and
kept there; for otherwise they will dry and shrink, and the peel
will not run. At first a person would find it no easy matter to
divest a rush of its peel or rind, so as to leave one regular,
narrow, even rib from top to bottom that may support the
pith: but this, like other feats, soon becomes familiar even to
children; and we have seen an old woman, stone-blind,
performing this business with great dispatch, and seldom
failing to strip them with the nicest regularity. When these
junci are thus far prepared, they must lie out on the grass to be

bleached, and take the dew for some nights, and afterwards be dried in the sun.

Some address is required in dipping these rushes in the scalding fat or grease; but this knack also is to be attained by practice. The careful wife of an industrious Hampshire labourer obtains all her fat for nothing; for she saves the scummings of her bacon-pot for this use; and, if the grease abounds with salt, she causes the salt to precipitate to the bottom, by setting the scummings in a warm oven. Where hogs are not much in use, and especially by the sea-side, the coarser animal-oils will come very cheap. A pound of common grease may be procured for four pence; and about six pounds of grease will dip a pound of rushes; and one pound of rushes may be bought for one shilling: so that a pound of rushes, medicated and ready for use, will cost three shillings. If men that keep bees will mix a little wax with the grease, it will give it a consistency, and render it more cleanly, and make the rushes burn longer: mutton-suet would have the same effect.

A good rush, which measured in length two feet four inches and an half, being minuted, burnt only three minutes short of an hour: and a rush still of greater length has been known to burn one hour and a quarter.

These rushes give a good clear light. Watch-lights (coated with tallow), it is true, shed a dismal one, 'darkness visible' [Milton, *Paradise Lost*, 1]; but then the wicks of those have two ribs of the rind, or peel, to support the pith, while the wick of the dipped rush has but one. The two ribs are intended to impede the progress of the flame and make the candle last.

In a pound of dry rushes, avoirdupois, which I caused to be weighed and numbered, we found upwards of one thousand six hundred individuals. Now, suppose each of these burns, one with another, only half an hour, then a poor man will purchase eight hundred hours of light, a time exceeding thirty-three entire days, for three shillings. According to this account each rush, before dipping, costs $\frac{1}{33}$ of a farthing, and $\frac{1}{11}$ afterwards. Thus a poor family will enjoy $5\frac{1}{2}$ hours of comfortable light for a farthing. An experienced old housekeeper assures me that one pound and an half of rushes completely supplies his family the year round,* since working people burn no candle in the long days, because they rise and go to bed by daylight.

Little farmers use rushes much in the short days, both morning and evening, in the dairy and kitchen; but the very poor, who are always the worst economists,* and therefore must continue very poor, buy an half-penny candle every evening, which, in their blowing open rooms, does not burn much more than two hours. Thus have they only two hours' light for their money instead of eleven.

While on the subject of rural economy, it may not be improper to mention a pretty implement of housewifery that we have seen no where else; that is, little neat besoms which our foresters make from the stalks of the *polytricum commune*, or great golden maiden-hair, which they call silk-wood, and find plenty in the bogs. When this moss is well combed and dressed, and divested of its outer skin, it becomes of a beautiful bright-chestnut colour; and, being soft and pliant, is very proper for the dusting of beds, curtains, carpets, hangings, &c. If these besoms were known to the brush-makers in town, it is probable they might come much in use for the purpose above-mentioned.[14]

I am, &c.

Letter 27

Selborne, Dec. 12, 1775.

DEAR SIR,

We had in this village more than twenty years ago an idiot-boy,* whom I well remember, who, from a child, shewed a strong propensity to bees; they were his food, his amusement, his sole object. And as people of this cast have seldom more than one point in view, so this lad exerted all his few faculties on this one pursuit. In the winter he dozed away his time, within his father's house, by the fireside, in a kind of torpid state, seldom departing from the chimney-corner; but in the summer he was all alert, and in quest of his game in the fields, and on sunny banks. Honey-bees, humble-bees, and wasps, were his prey wherever he found them: he had no apprehensions from their stings, but would seize them *nudis manibus* [with bare hands], and at once disarm them of their weapons, and suck their bodies for the sake of their honey-bags. Sometimes he would fill his bosom between his shirt and his skin with a number of these captives; and sometimes would confine them in bottles. He was a very *merops apiaster*, or bee-bird; and very injurious to men that kept bees; for he would slide into their bee-gardens, and, sitting down before the stools,* would rap with his finger on the hives, and so take the bees as they came out. He has been known to overturn hives for the sake of honey, of which he was passionately fond. Where metheglin was making he would linger round the tubs and vessels, begging a draught of what he called bee-wine. As he ran about he used to make a humming noise with his lips, resembling the buzzing of bees. This lad was lean and sallow, and of a cadaverous complexion; and, except in his favourite pursuit, in which he was wonderfully adroit, discovered no manner of understanding. Had his capacity been better, and directed to the same object, he had perhaps abated much of our wonder at the feats of a more modern exhibiter of bees: and we may justly say of him now,

> Thou,*
> Had thy presiding star propitious shone,
> Should'st *Wildman* be . . .

When a tall youth he was removed from hence to a distant village, where he died, as I understand, before he arrived at manhood.

I am, &c.

Letter 28

Selborne, Jan. 8, 1776.

DEAR SIR,

It is the hardest thing in the world to shake off superstitious prejudices: they are sucked in as it were with our mother's milk; and, growing up with us at a time when they take the fastest hold and make the most lasting impressions, become so interwoven into our very constitutions, that the strongest good sense is required to disengage ourselves from them. No wonder therefore that the lower people retain them their whole lives through, since their minds are not invigorated by a liberal education, and therefore not enabled to make any efforts adequate to the occasion.

Such a preamble seems to be necessary before we enter on the superstitions of this district, lest we should be suspected of exaggeration in a recital of practices too gross for this enlightened age.

But the people of Tring, in Hertfordshire, would do well to remember, that no longer ago than the year 1751, and within twenty miles of the capital, they seized on two superannuated wretches, crazed with age, and overwhelmed with infirmities, on a suspicion of witchcraft; and, by trying experiments, drowned them in a horse-pond.

In a farm-yard near the middle of this village stands, at this day, a row of pollard-ashes, which, by the seams and long cicatrices down their sides, manifestly shew that, in former times, they have been cleft asunder. These trees, when young and flexible, were severed and held open by wedges, while ruptured children, stripped naked, were pushed through the apertures, under a persuasion that, by such a process, the poor babes would be cured of their infirmity. As soon as the operation was over, the tree, in the suffering part, was plastered with loam, and carefully swathed up. If the parts coalesced and soldered together, as usually fell out, where the feat was performed with any adroitness at all, the party was cured; but, where the cleft continued to gape, the operation, it was supposed, would prove ineffectual. Having occasion to enlarge my garden not long since, I cut down two or three such trees, one of which did not grow together.*

We have several persons now living in the village, who, in their childhood, were supposed to be healed by this superstitious ceremony, derived down perhaps from our Saxon ancestors, who practised it before their conversion to Christianity.

At the south corner of the Plestor, or area, near the church, there stood, about twenty-years ago, a very old grotesque hollow pollard-ash, which for ages had been looked on with no small veneration as a shrew-ash. Now a shrew-ash is an ash whose twigs or branches, when gently applied to the limbs of cattle, will immediately relieve the pains which a beast suffers from the running of a shrew-mouse over the part affected: for it is supposed that a shrew-mouse is of so baneful and deleterious a nature, that wherever it creeps over a beast, be it horse, cow, or sheep, the suffering animal is afflicted with cruel anguish, and threatened with the loss of the use of the limb. Against this accident, to which they were continually liable, our provident fore-fathers always kept a shrew-ash at hand, which, when once medicated, would maintain its virtue for ever. A shrew-ash was made thus:[15]—into the body of the tree a deep hole was bored with an auger, and a poor devoted shrew-mouse was thrust in alive, and plugged in, no doubt, with several quaint incantations long since forgotten. As the ceremonies necessary for such a consecration are no longer understood, all succession is at an end, and no such tree is known to exist in the manor, or hundred.

As to that on the Plestor

> The late vicar stubb'd and burnt it*

when he was way-warden, regardless of the remonstrances of the by-standers, who interceded in vain for its preservation, urging its power and efficacy, and alleging that it had been

> Religione patrum multos servata per annos
>
> [Saved for many years by the reverence of
>   our ancestors.   Virgil, *Aeneid* 2]

I am, &c.

Letter 29

Selborne, Feb. 7, 1776.

DEAR SIR,

In heavy fogs, on elevated situations especially, trees are perfect alembics; and no one that has not attended to such matters can imagine how much water one tree will distil in a night's time, by condensing the vapour, which trickles down the twigs and boughs, so as to make the ground below quite in a float. In Newton-lane, in October 1775, on a misty day, a particular oak in leaf dropped so fast that the cart-way stood in puddles and the ruts ran with water, though the ground in general was dusty.*

In some of our smaller islands in the West-Indies, if I mistake not, there are no springs or rivers; but the people are supplied with that necessary element, water, merely by the dripping of some large tall trees, which, standing in the bosom of a mountain, keep their heads constantly enveloped with fogs and clouds, from which they dispense their kindly never-ceasing moisture; and so render those districts habitable by condensation alone.

Trees in leaf have such a vast proportion more of surface than those that are naked, that, in theory, their condensations should greatly exceed those that are stripped of their leaves; but, as the former imbibe also a great quantity of moisture, it is difficult to say which drip most: but this I know, that deciduous trees that are entwined with much ivy seem to distil the greatest quantity. Ivy-leaves are smooth, and thick, and cold, and therefore condense very fast; and besides ever-greens imbibe very little. These facts may furnish the intelligent with hints concerning what sorts of trees they should plant round small ponds that they would wish to be perennial; and shew them how advantageous some trees are in preference to others.

Trees perspire profusely, condense largely, and check evaporation so much, that woods are always moist; no wonder therefore that they contribute much to pools and streams.

That trees are great promoters of lakes and rivers appears from a well known fact in North-America; for, since the woods and forests have been grubbed and cleared, all bodies of water are much diminished; so that some streams, that were very considerable a century ago, will not now drive a common mill.[16] Besides, most woodlands, forests, and chases, with us abound with pools and morasses; no doubt for the reason given above.

To a thinking mind few phenomena are more strange than the state of little ponds on the summits of chalk-hills, many of which are never dry in the most trying droughts of summer. On chalk-hills I say, because in many rocky and gravelly soils springs usually break out pretty high on the sides of elevated grounds and mountains; but no person acquainted with chalky districts will allow that they ever saw springs in such a soil but in vallies and bottoms, since the waters of so pervious a stratum as chalk all lie on one dead level, as well-diggers have assured me again and again.

Now we have many such little round ponds in this district; and one in particular on our sheep-down, three hundred feet above my house; which, though never above three feet deep in the middle, and not more than thirty feet in diameter, and containing perhaps not more than two or three hundred hogsheads of water, yet never is known to fail, though it affords drink for three hundred or four hundred sheep, and for at least twenty head of large cattle beside. This pond, it is true, is over-hung with two moderate beeches, that, doubtless, at times afford it much supply: but then we have others as small, that, without the aid of trees, and in spite of evaporation from sun and wind, and perpetual consumption by cattle, yet constantly maintain a moderate share of water, without over-flowing in the wettest seasons, as they would do if supplied by springs. By my journal of May, 1775, it appears that 'the small and even considerable ponds in the vales are now dried up, while the small ponds on the very tops of hills are but little affected'.* Can this difference be accounted for from evaporation alone, which certainly is more prevalent in bottoms? or rather have not those elevated pools some unnoticed recruits, which in the night time counterbalance the waste of the day; without which the cattle alone must soon exhaust them? And here it will be necessary to enter more minutely into the cause. Dr Hales, in his *Vegetable Statics*, advances, from experiment, that 'the moister the earth is the more dew falls on it in a night: and more than a double quantity of dew falls on a surface of water than there does on an equal surface of moist earth.' Hence we see that water, by its coolness, is enabled to assimilate to itself a large quantity of moisture nightly by condensation; and that the air, when loaded with fogs and vapours, and even with copious dews, can alone advance a considerable and never-failing resource.* Persons that are much abroad, and travel early and late; such as shepherds, fishermen, &c. can tell what prodigious fogs prevail in the night on elevated downs, even in the hottest parts of summer; and how much the surfaces of things are drenched by those swimming vapours, though, to the senses, all the while, little moisture seems to fall.

I am, &c.

Letter 30

Selborne, April 3, 1776.

DEAR SIR,

Monsieur Herissant, a French anatomist, seems persuaded that he has discovered the reason why cuckoos do not hatch their own eggs; the impediment, he supposes, arises from the internal structure of their parts, which incapacitates them for incubation. According to this gentleman, the crop, or craw, of a cuckoo does not lie before the sternum at the bottom of the neck, as in the *gallinae*, *columbae*, &c. but immediately behind it, on and over the bowels, so as to make a large protuberance in the belly.[17]

Induced by this assertion, we procured a cuckoo; and, cutting open the breast-bone, and exposing the intestines to sight, found the crop lying as mentioned above. The stomach was large and round, and stuffed hard like a pincushion with food, which, upon nice examination, we found to consist of various insects; such as small scarabs, spiders, and dragon-flies; the last of which we have seen cuckoos catching on the wing as they were just emerging out of the aurelia state. Among this farrago also were to be seen maggots, and many seeds, which belonged either to gooseberries, currants, cran-berries, or some such fruit; so that these birds apparently subsist on insects and fruits: nor was there the least appearance of bones, feathers, or fur, to support the idle notion of their being birds of prey.

The sternum in this bird seemed to us to be remarkably

short, between which and the anus lay the crop, or craw, and immediately behind that the bowels against the back-bone.

It must be allowed, as this anatomist observes, that the crop placed just upon the bowels must, especially when full, be in a very uneasy situation during the business of incubation; yet the test will be to examine whether birds that are actually known to sit for certain are not formed in a similar manner. This inquiry I proposed to myself to make with a fern-owl, or goat-sucker, as soon as opportunity offered: because, if their information proves the same, the reason for incapacity in the cuckoo will be allowed to have been taken up somewhat hastily.

Not long after a fern-owl was procured, which, from its habits and shape, we suspected might resemble the cuckoo in its internal construction. Nor were our suspicions ill-grounded; for, upon the dissection, the crop, or craw, also lay behind the sternum, immediately on the viscera, between them and the skin of the belly. It was bulky, and stuffed hard with large *phalaenae*, moths of several sorts, and their eggs, which no doubt had been forced out of these insects by the action of swallowing.*

Now as it appears that this bird, which is so well known to practise incubation, is formed in a similar manner with cuckoos, Monsieur Herissant's conjecture, that cuckoos are incapable of incubation from the disposition of their intestines, seems to fall to the ground; and we are still at a loss for the cause of that strange and singular peculiarity in the instance of the *cuculus canorus*.

We found the case to be the same with the ring-tail hawk, in respect to formation; and, as far as I can recollect, with the swift; and probably it is so with many more sorts of birds that are not granivorous.

I am, &c.

Letter 31

Selborne, April 29, 1776.

DEAR SIR,

On August the 4th, 1775, we surprised a large viper, which seemed very heavy and bloated, as it lay in the grass basking in the sun.* When we came to cut it up, we found that the abdomen was crowded with young, fifteen in number; the shortest of which measured full seven inches, and were about the size of full-grown earth-worms. This little fry issued into the world with the true viper-spirit about them, shewing great alertness as soon as disengaged from the belly of the dam: they twisted and wriggled about, and set themselves up and gaped very wide when touched with a stick, shewing manifest tokens of menace and defiance, though as yet they had no manner of fangs that we could find, even with the help of our glasses.

To a thinking mind nothing is more wonderful than that early instinct which impresses young animals with the notion of the situation of their natural weapons, and of using them properly in their own defence, even before those weapons subsist or are formed. Thus a young cock will spar at his adversary before his spurs are grown; and a calf or lamb will push with their heads before their horns are sprouted. In the same manner did these young adders attempt to bite before their fangs were in being. The dam however was furnished with very formidable ones, which we lifted up (for they fold down when not used) and cut them off with the point of our scissars.

There was little room to suppose that this brood had ever been in the open air before; and that they were taken in for refuge, at the mouth of the dam, when she perceived that danger was approaching; because then probably we should have found them somewhere in the neck, and not in the abdomen.

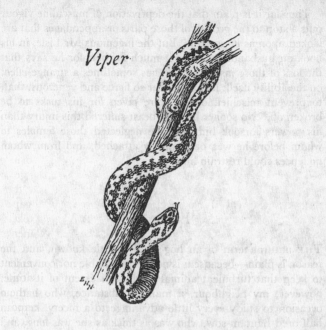

Viper

Letter 32

CASTRATION has a strange effect: it emasculates both man, beast, and bird, and brings them to a near resemblance of the other sex. Thus eunuchs have smooth unmuscular arms, thighs, and legs; and broad hips, and beardless chins, and squeaking voices. Gelt-stags and bucks have hornless heads, like hinds and does. Thus wethers have small horns, like ewes; and oxen large bent horns, and hoarse voices when they low, like cows: for bulls have short straight horns; and though they mutter and grumble in a deep tremendous tone, yet they low in a shrill high key. Capons have small combs and gills, and look pallid about the head, like pullets; they also walk without any parade, and hover chickens like hens. Barrow-hogs have also small tusks like sows.

Thus far it is plain that the deprivation of masculine vigour puts a stop to the growth of those parts or appendages that are looked upon as its insignia. But the ingenious Mr Lisle, in his book on husbandry, carries it much farther; for he says that the loss of those insignia alone has sometimes a strange effect on the ability itself: he had a boar so fierce and venerous, that, to prevent mischief, orders were given for his tusks to be broken off.* No sooner had the beast suffered this injury than his powers forsook him, and he neglected those females to whom before he was passionately attached, and from whom no fences could restrain him.

Letter 33

THE natural term of an hog's life is little known, and the reason is plain—because it is neither profitable nor convenient to keep that turbulent animal to the full extent of its time: however, my neighbour, a man of substance, who had no occasion to study every little advantage to a nicety, kept an half bred Bantam-sow, who was as thick as she was long, and whose belly swept on the ground till she was advanced to her seventeenth year; at which period she shewed some tokens of age by the decay of her teeth and the decline of her fertility.

For about ten years this prolific mother produced two litters in the year of about ten at a time, and once above twenty at a litter; but, as there were near double the number of pigs to that of teats, many died. From long experience in the world, this female was grown very sagacious and artful:—when she found occasion to converse with a boar she used to open all the intervening gates, and march, by herself, up to a distant farm where one was kept; and when her purpose was served would return by the same means. At the age of about fifteen her litters began to be reduced to four or five; and such a litter she exhibited when in her fatting-pen. She proved, when fat, good bacon, juicy, and tender; the rind, or sward, was remarkably thin. At a moderate computation she was allowed to have been the fruitful parent of three hundred pigs: a prodigious instance of fecundity in so large a quadruped!* She was killed in spring 1775.

I am, &c.

Letter 34

Selborne, May 9, 1776.

DEAR SIR,

. . . admorunt ubera tigres

[tigresses (must have) suckled
(you).   Virgil, *Aeneid* 4]

We have remarked in a former letter how much incongruous animals, in a lonely state, may be attached to each other from a spirit of sociality; in this it may not be amiss to recount a different motive which has been known to create as strange a fondness.

My friend had a little helpless leveret brought to him, which the servants fed with milk in a spoon, and about the same time his cat kittened and the young were dispatched and buried. The hare was soon lost, and supposed to be gone the way of most fondlings, to be killed by some dog or cat. However, in about a fortnight, as the master was sitting in his garden in the dusk of the evening, he observed his cat, with tail erect, trotting towards him, and calling with little short inward notes of complacency, such as they use towards their kittens, and something gamboling after, which proved to be the leveret that the cat had supported with her milk, and continued to support with great affection.

Thus was a graminivorous animal nurtured by a carnivorous and predaceous one!

Why so cruel and sanguinary a beast as a cat, of the ferocious genus of *Feles*, the *murium leo*, as Linnaeus calls it, should be affected with any tenderness towards an animal which is its natural prey, is not so easy to determine.

This strange affection probably was occasioned by that desiderium, those tender maternal feelings, which the loss of her kittens had awakened in her breast; and by the complacency and ease she derived to herself from the procuring her teats to be drawn, which were too much distended with milk, till, from habit, she became as much delighted with this foundling as if it had been her real offspring.*

This incident is no bad solution of that strange circumstance which grave historians as well as the poets assert, of exposed children being sometimes nurtured by female wild beasts that probably had lost their young. For it is not one whit more marvellous that Romulus and Remus, in their infant state, should be nursed by a she-wolf, than that a poor little sucking leveret should be fostered and cherished by a bloody grimalkin.

> . . . viridi foetam Mavortis in antro
> Procubuisse lupam: geminus huic ubera circum
> Ludere pendentes pueros, et lambere matrem
> Impavidos: illam tereti cervice reflexam
> Mulcere alternos, et corpora fingere lingua.

[He, that is Vulcan, had depicted 'the newly delivered she-wolf lying outstretched in the green cave of Mars, the twin boys playing as they hung around her breasts licking their dam without fear, while she, bending back her shapely neck, caressed them by turns and moulded their bodies with her tongue'.   Virgil, *Aeneid* 8]

Letter 35

Selborne, May 20, 1777.

DEAR SIR,

Lands that are subject to frequent inundations are always poor; and probably the reason may be because the worms are drowned. The most insignificant insects and reptiles are of much more consequence, and have much more influence in the economy of Nature, than the incurious are aware of; and are mighty in their effect, from their minuteness, which renders them less an object of attention; and from their numbers and fecundity. Earth-worms, though in appearance a small and despicable link in the chain of Nature, yet, if lost, would make a lamentable chasm. For, to say nothing of half the birds, and some quadrupeds which are almost entirely supported by them, worms seem to be the great promoters of vegetation, which would proceed but lamely without them, by boring, perforating, and loosening the soil, and rendering it pervious to rains and the fibres of plants, by drawing straws and stalks of leaves and twigs into it; and, most of all, by throwing up such infinite numbers of lumps of earth called wormcasts, which, being their excrement, is a fine manure for grain and grass. Worms probably provide new soil for hills and slopes where the rain washes the earth away; and they affect slopes, probably to avoid being flooded. Gardeners and farmers express their detestation of worms; the former because they render their walks unsightly, and make them much work: and the latter because, as they think, worms eat their green corn. But these men would find that the earth without worms would soon become cold, hard-bound, and void of fermentation; and consequently steril: and besides, in favour of worms, it should be hinted that green corn, plants, and flowers, are not so much injured by them as by many species of *coleoptera* (scarabs), and *tipulae* (long-legs), in their larva, or grub-state; and by unnoticed myriads of small shell-less snails, called slugs, which silently and imperceptibly make amazing havoc in the field and garden.[18]

These hints we think proper to throw out in order to set the inquisitive and discerning to work.

A good monography of worms would afford much entertainment and information at the same time, and would open a large and new field in natural history. Worms work most in the spring; but by no means lie torpid in the dead months; are out every mild night in the winter,* as any person may be convinced that will take the pains to examine his grass-plots with a candle; are hermaphrodites, and much addicted to venery, and consequently very prolific.

I am, &c.

Letter 36

Selborne, Nov. 22, 1777.

DEAR SIR,

You cannot but remember that the twenty-sixth and twenty-seventh of last March were very hot days; so sultry that every body complained and were restless under those sensations to which they had not been reconciled by gradual approaches.

This sudden summer-like heat was attended by many summer coincidences; for on those two days the thermometer rose to sixty-six in the shade; many species of insects revived and came forth; some bees swarmed in this neighbourhood;* the old tortoise, near Lewes, in Sussex, awakened and came forth out of its dormitory; and, what is most to my present purpose, many house-swallows appeared and were very alert in many places, and particularly at Cobham, in Surrey.

But as that short warm period was succeeded as well as preceded by harsh severe weather, with frequent frosts and ice, and cutting winds, the insects withdrew, the tortoise retired again into the ground, and the swallows were seen no more until the tenth of April, when, the rigour of the spring abating, a softer season began to prevail.

Again; it appears by my journals for many years past that house-martins retire, to a bird, about the beginning of October; so that a person not very observant of such matters would conclude that they had taken their last farewell: but then it may be seen in my diaries also that considerable flocks have discovered themselves again in the first week of November, and often on the fourth day of that month only for one day; and that not as if they were in actual migration, but playing about at their leisure and feeding calmly, as if no enterprize of moment at all agitated their spirits. And this was the case in the beginning of this very month; for, on the fourth of November, more than twenty house-martins, which, in appearance, had all departed about the seventh of October, were seen again, for that one morning only, sporting between my fields and the Hanger, and feasting on insects which swarmed in that sheltered district. The preceding day was wet and blustering, but the fourth was dark and mild, and soft, the wind at south-west, and the thermometer at $58\frac{1}{2}°$; a pitch not common at that season of the year. Moreover, it may not be amiss to add in this place, that whenever the thermometer is above the $50°$ the bat comes flitting out in every autumnal and winter-month.

From all these circumstances laid together, it is obvious that torpid insects, reptiles, and quadrupeds, are awakened from their profoundest slumbers by a little untimely warmth; and therefore that nothing so much promotes this death-like stupor as a defect of heat.* And farther, it is reasonable to suppose that two whole species, or at least many individuals of those two species, of British hirundines, do never leave this island at all, but partake of the same benumbed state: for we cannot suppose that, after a month's absence, house-martins can return from southern regions to appear for one morning in November, or that house-swallows should leave the districts of Africa to enjoy, in March, the transient summer of a couple of days.

I am, &c.

Letter 37

Selborne, Jan. 8, 1778.

DEAR SIR,

There was in this little village several years ago a miserable pauper, who, from his birth, was afflicted with a leprosy, as far as we are aware of a singular kind, since it affected only the palms of his hands and the soles of his feet. This scaly eruption usually broke out twice in the year, at the spring and fall; and, by peeling away, left the skin so thin and tender that neither his hands or feet were able to perform their functions; so that the poor object was half his time on crutches, incapable of employ, and languishing in a tiresome state of indolence and inactivity. His habit was lean, lank, and cadaverous. In this sad plight he dragged on a miserable existence, a burden to himself and his parish, which was obliged to support him till he was relieved by death at more than thirty years of age.

The good women, who love to account for every defect in

children by the doctrine of longing, said that his mother felt a violent propensity for oysters, which she was unable to gratify; and that the black rough scurf on his hands and feet were the shells of that fish. We knew his parents, neither of which were lepers; his father in particular lived to be far advanced in years.

In all ages the leprosy has made dreadful havock among mankind. The Israelites seem to have been greatly afflicted with it from the most remote times; as appears from the peculiar and repeated injunctions given them in the Levitical law.[19] Nor was the rancour of this foul disorder much abated in the last period of their commonwealth, as may be seen in many passages of the New Testament.

Some centuries ago this horrible distemper prevailed all Europe over; and our forefathers were by no means exempt, as appears by the large provision made for objects labouring under this calamity. There was an hospital for female lepers in the diocese of Lincoln, a noble one near Durham, three in London and Southwark, and perhaps many more in or near our great towns and cities. Moreover, some crowned heads, and other wealthy and charitable personages, bequeathed large legacies to such poor people as languished under this hopeless infirmity.

It must therefore, in these days, be, to an humane and thinking person, a matter of equal wonder and satisfaction, when he contemplates how nearly this pest is eradicated, and observes that a leper is now a rare sight. He will, moreover, when engaged in such a train of thought, naturally inquire for the reason. This happy change perhaps may have originated and been continued from the much smaller quantity of salted meat and fish now eaten in these kingdoms; from the use of linen next the skin; from the plenty of better bread; and from the profusion of fruits, roots, legumes, and greens, so common in every family. Three or four centuries ago, before there were any enclosures, sown-grasses, field-turnips, or field-carrots, or hay, all the cattle which had grown fat in summer, and were not killed for winter-use, were turned out soon after Michaelmas to shift as they could through the dead months; so that no fresh meat could be had in winter or spring. Hence the marvellous account of the vast stores of salted flesh found in the larder of the eldest Spencer[20] in the days of Edward the Second, even so late in the spring as the third of May. It was from magazines like these that the turbulent barons supported in idleness their riotous swarms of retainers ready for any disorder or mischief. But agriculture is now arrived at such a pitch of perfection, that our best and fattest meats are killed in the winter; and no man need eat salted flesh, unless he prefers it, that has money to buy fresh.

One cause of this distemper might be, no doubt, the quantity of wretched flesh and salt fish consumed by the commonalty at all seasons as well as in lent; which our poor now would hardly be persuaded to touch.*

The use of linen changes, shirts or shifts, in the room of sordid and filthy woollen, long worn next the skin, is a matter of neatness comparatively modern; but must prove a great means of preventing cutaneous ails. At this very time woollen instead of linen prevails among the poorer Welch, who are subject to foul eruptions.

The plenty of good wheaten bread that now is found among all ranks of people in the south, instead of that miserable sort which used in old days to be made of barley or beans, may contribute not a little to the sweetening their blood and correcting their juices; for the inhabitants of mountainous districts, to this day, are still liable to the itch and other cutaneous disorders, from a wretchedness and poverty of diet.

As to the produce of a garden, every middle-aged person of observation may perceive, within his own memory, both in town and country, how vastly the consumption of vegetables is increased. Green-stalls in cities now support multitudes in a comfortable state, while gardeners get fortunes. Every decent labourer also has his garden, which is half his support, as well as his delight; and common farmers provide plenty of beans, peas, and greens for their hinds to eat with their bacon; and those few that do not are despised for their sordid parsimony, and looked upon as regardless of the welfare of their dependants. Potatoes have prevailed in this little district, by means of premiums, within these twenty years only, and are much esteemed here now by the poor, who would scarce have ventured to taste them in the last reign.*

Our Saxon ancestors certainly had some sort of cabbage, because they call the month of February *sprout-cale*; but, long after their days, the cultivation of gardens was little attended to. The religious, being men of leisure, and keeping up a constant correspondence with Italy, were the first people among us that had gardens and fruit-trees in any perfection, within the walls of their abbies[21] and priories. The barons neglected every pursuit that did not lead to war or tend to the pleasure of the chase.

It was not till gentlemen took up the study of horticulture themselves that the knowledge of gardening made such hasty advances. Lord Cobham, Lord Ila, and Mr Waller of Beaconsfield,* were some of the first people of rank that promoted the elegant science of ornamenting without despising the superintendence of the kitchen quarters and fruit walls.

A remark made by the excellent Mr Ray in his Tour of Europe at once surprises us, and corroborates what has been advanced above; for we find him observing, so late as his days, that 'the Italians use several herbs for sallets, which are not yet or have not been but lately used in England, viz. *selleri* (celery) which is nothing else but the sweet smallage; the young shoots whereof, with a little of the head of the root cut off, they eat raw with oil and pepper.' And farther he adds 'curled endive blanched is much used beyond seas; and, for a raw sallet, seemed to excel lettuce itself.' Now this journey was undertaken no longer ago than in the year 1663.

I am, &c.

Letter 38

Forte puer, comitum seductus ab agmine fido,
Dixerat, ecquis adest? et, adest, responderat echo.
Hic stupet; utque aciem partes divisit in omnes;
Voce, veni, clamat magna. Vocat illa vocantem.

[Perchance a boy, parted from his trusty band of comrades, had exclaimed, 'Is anyone here?' to which had come an echoing response, 'Anyone here'. In amazement, and turning his keen glance in all directions, the boy calls out in a loud voice, 'Come!' And his voice is answered as he calls.   Ovid, *Metamorphoses* 3]

Selborne, Feb. 12, 1778.

DEAR SIR,

In a district so diversified as this, so full of hollow vales and hanging woods, it is no wonder that echoes should abound. Many we have discovered that return the cry of a pack of dogs, the notes of a hunting-horn, a tunable ring of bells, or the melody of birds, very agreeably: but we were still at a loss for a polysyllabical, articulate echo, till a young gentleman, who had parted from his company in a summer evening walk, and was calling after them, stumbled upon a very curious one in a spot where it might least be expected. At first he was much surprised, and could not be persuaded but that he was mocked by some boy; but, repeating his trials in several languages, and finding his respondent to be a very adroit polyglot, he then discerned the deception.

This echo, in an evening, before rural noises cease, would repeat ten syllables most articulately and distinctly, especially if quick dactyls were chosen. The last syllables of

Tityre, tu patulae recubans . . .

[You, O Tityrus, reclining (beneath
the canopy of your) spreading
(beech).   Virgil, *Eclogues* 1]

were as audibly and intelligibly returned as the first: and there is no doubt, could trial have been made, but that at midnight, when the air is very elastic, and a dead stillness prevails, one or two syllables more might have been obtained; but the distance rendered so late an experiment very inconvenient.

Quick dactyls, we observed, succeeded best; for when we came to try its powers in slow, heavy, embarrassed spondees of the same number of syllables,

Monstrum horrendum, informe, ingens . . .

[A monster, horrible, misshapen, huge . . .
                    Virgil, *Aeneid* 3]

we could perceive a return but of four or five.

All echoes have some one place to which they are returned stronger and more distinct than to any other; and that is always the place that lies at right angles with the object of repercussion, and is not too near, nor too far off. Buildings, or naked rocks, re-echo much more articulately than hanging wood or vales; because in the latter the voice is as it were entangled, and embarrassed in the covert, and weakened in the rebound.

The true object of this echo, as we found by various experiments, is the stone-built, tiled hop-kiln in Gally-lane, which measures in front 40 feet, and from the ground to the eaves 12 feet. The true *centrum phonicum*, or just distance, is one particular spot in the King's-field, in the path to Nore-hill, on the very brink of the steep balk above the hollow cart way. In this case there is no choice of distance; but the path, by meer contingency, happens to be the lucky, the identical spot, because the ground rises or falls so immediately, if the speaker either retires or advances, that his mouth would at once be above or below the object.

We measured this polysyllabical echo with great exactness, and found the distance to fall very short of Dr Plot's rule for distant articulation: for the Doctor, in his history of *Oxfordshire*, allows 120 feet for the return of each syllable distinctly: hence this echo, which gives ten distinct syllables, ought to measure 400 yards, or 120 feet to each syllable; whereas our distance is only 258 yards, or near 75 feet, to each syllable. Thus our measure falls short of the Doctor's, as five to eight: but then it must be acknowledged that this candid philosopher was convinced afterwards, that some latitude must be admitted of in the distance of echoes according to time and place.

When experiments of this sort are making, it should always be remembered that weather and the time of day have a vast influence on an echo; for a dull, heavy, moist air deadens and clogs the sound; and hot sunshine renders the air thin and weak, and deprives it of all its springiness; and a ruffling wind quite defeats the whole. In a still, clear, dewy evening the air is most elastic; and perhaps the later the hour the more so.

Echo has always been so amusing to the imagination, that

the poets have personified her; and in their hands she has been the occasion of many a beautiful fiction. Nor need the gravest man be ashamed to appear taken with such a phaenomenon, since it may become the subject of philosophical or mathematical inquiries.

One should have imagined that echoes, if not entertaining, must at least have been harmless and inoffensive; yet Virgil advances a strange notion, that they are injurious to bees. After enumerating some probable and reasonable annoyances, such as prudent owners would wish far removed from their bee-gardens, he adds

> . . . aut ubi concava pulsu
> Saxa sonant, vocisque offensa resultat imago
>
> [or where the arching rocks vibrate when struck,
> and the sound, thrown back, rebounds in an
> echo. Virgil, *Georgics* 4]

This wild and fanciful assertion will hardly be admitted by the philosophers of these days; especially as they all now seem agreed that insects are not furnished with any organs of hearing at all. But if it should be urged, that though they cannot *hear* yet perhaps they may *feel* the repercussions of sounds, I grant it is possible they may. Yet that these impressions are distasteful or hurtful, I deny, because bees, in good summers, thrive well in my outlet, where the echoes are very strong: for this village is another Anathoth,* a place of responses or echoes. Besides, it does not appear from experiment that bees are in any way capable of being affected by sounds: for I have often tried my own with a large speaking-trumpet held close to their hives, and with such exertion of voice as would have hailed a ship at the distance of a mile, and still these insects pursued their various employments undisturbed, and without shewing the least sensibility or resentment.

Some time since its discovery this echo is become totally silent, though the object, or hop-kiln, remains: nor is there any mystery in this defect; for the field between is planted as an hop-garden, and the voice of the speaker is totally absorbed and lost among the poles and entangled foliage of the hops.

And when the poles are removed in autumn the disappointment is the same; because a tall quick-set hedge, nurtured up for the purpose of shelter to the hop ground, entirely interrupts the impulse and repercussion of the voice: so that till those obstructions are removed no more of its garrulity can be expected.

Should any gentleman of fortune think an echo in his park or outlet a pleasant incident, he might build one at little or no expense.* For whenever he had occasion for a new barn, stable, dog-kennel, or the like structure, it would be only needful to erect this building on the gentle declivity of an hill, with a like rising opposite to it, at a few hundred yards distance; and perhaps success might be the easier ensured could some canal, lake, or stream, intervene. From a seat at the *centrum phonicum* he and his friends might amuse themselves sometimes of an evening with the prattle of this loquacious nymph; of whose complacency and decent reserve more may be said than can with truth of every individual of her sex; since she is

> . . . quae nec reticere loquenti,
> Nec prior ipsa loqui didicit resonabilis echo.

> [. . . resounding Echo, who could neither hold
>   her peace when others spoke, nor begin to
>   speak until others had spoken to her.
>   Ovid, *Metamorphoses* 3]

I am, &c.

P.S. The classic reader will, I trust, pardon the following lovely quotation, so finely describing echoes, and so poetically accounting for their causes from popular superstition:

> Quae bene quom videas, rationem reddere possis
> Tute tibi atque aliis, quo pacto per loca sola
> Saxa pareis formas verborum ex ordine reddant,
> Palanteis comites quom monteis inter opacos
> Quaerimus, et magna dispersos voce ciemus.
> Sex etiam, aut septem loca vidi reddere voces
> Unam quom jaceres: ita colles collibus ipsis
> Verba repulsantes iterabant dicta referre.
> Haec loca capripedes Satyros, Nymphasque tenere

Finitimi fingunt, et Faunos esse loquuntur;
Quorum noctivago strepitu, ludoque jocanti
Adfirmant volgo taciturna silentia rumpi,
Chordarumque sonos fieri, dulceisque querelas,
Tibia quas fundit digitis pulsata canentum;
Et genus agricolum late sentiscere, quom Pan
Pinea semiferi capitis velamina quassans,
Unco saepe labro calamos percurrit hianteis,
Fistula silvestrem ne cesset fundere musam.

(Lucretius, *Lib.* iv)*

[When you perceive this well, you may be able to give a reason to yourself and others, how it is in solitary places that the rocks give back the same shapes of words in their order, when we seek straying comrades amongst the shady mountains and call loudly upon them to all sides. I have even seen places give back six or seven cries, when you uttered one: so did hill to hill themselves buffet back and repeat the words thus trained to come back.

Such places the neighbours imagine to be haunted by goatfoot satyrs and nymphs, and they say there are fauns, by whose night-wandering noise and jocund play they commonly declare the voiceless silence to be broken, with the sound of strings and sweet plaintive notes, which the pipe sends forth touched by the player's fingers; they tell how the farmers' men all over the countryside listen, while Pan, shaking the pine leaves that cover his half-human head, often runs over the open reeds with curved lips, that the panpipes may never slacken in their flood of woodland music.]

Letter 39

Selborne, May 13, 1778.

DEAR SIR,

Among the many singularities attending those amusing birds the swifts, I am now confirmed in the opinion that we have every year the same number of pairs invariably; at least the result of my inquiry has been exactly the same for a long time past. The swallows and martins are so numerous, and so widely distributed over the village, that it is hardly possible to recount them; while the swifts, though they do not all build in the church, yet so frequently haunt it, and play and rendezvous round it, that they are easily enumerated. The number that I constantly find are eight pairs; about half of which reside in the church, and the rest build in some of the lowest and meanest thatched cottages. Now as these eight pairs, allowance being made for accidents, breed yearly eight pairs more, what becomes annually of this increase; and what determines every spring which pairs shall visit us, and re-occupy their ancient haunts?

Ever since I have attended to the subject of ornithology, I have always supposed that the sudden reverse of affection, that strange ἀντιστοργή, which immediately succeeds in the feathered kind to the most passionate fondness, is the occasion of an equal dispersion of birds over the face of the earth. Without this provision one favourite district would be crowded with inhabitants, while others would be destitute and forsaken. But the parent birds seem to maintain a jealous superiority, and to oblige the young to seek for new abodes: and the rivalry of the males, in many kinds, prevents their crowding the one on the other. Whether the swallows and house-martins return in the same exact number annually is not easy to say, for reasons given above: but it is apparent, as I have remarked before in my Monographies, that the numbers returning bear no manner of proportion to the numbers retiring.*

Letter 40

Selborne, June 2, 1778.

DEAR SIR,

The standing objection to botany has always been, that it is a pursuit that amuses the fancy and exercises the memory, without improving the mind, or advancing any real knowledge: and, where the science is carried no farther than a mere systematic classification, the charge is but too true. But the botanist that is desirous of wiping off this aspersion should be by no means content with a list of names; he should study plants philosophically, should investigate the laws of vegetation, should examine the powers and virtues of efficacious herbs, should promote their cultivation; and graft the gardener, the planter, and the husbandman, on the phytologist. Not that system is by any means to be thrown aside; without system the field of Nature would be a pathless wilderness; but system should be subservient to, not the main object of, pursuit.

Vegetation is highly worthy of our attention, and in itself is of the utmost consequence to mankind, and productive of many of the greatest comforts and elegancies of life. To plants we owe timber, bread, beer, honey, wine, oil, linen, cotton, &c. what not only strengthens our hearts, and exhilarates our spirits, but what secures us from inclemencies of weather and adorns our persons. Man, in his true state of nature, seems to be subsisted by spontaneous vegetation: in middle climes, where grasses prevail, he mixes some animal food with the produce of the field and garden: and it is towards the polar extremes only that, like his kindred bears and wolves, he gorges himself with flesh alone, and is driven, to what hunger has never been known to compel the very beasts, to prey on his own species.[22]

The productions of vegetation have had a vast influence on the commerce of nations, and have been the great promoters of navigation, as may be seen in the articles of sugar, tea, tobacco, opium, ginseng, betel, pepper, &c. As every climate has its peculiar produce, our natural wants bring on a mutual intercourse; so that by means of trade each distant part is supplied with the growth of every latitude. But, without the knowledge of plants and their culture, we must have been content with our hips and haws, without enjoying the delicate fruits of India and the salutiferous drugs of Peru.

Instead of examining the minute distinctions of every various species of each obscure genus, the botanist should endeavour to make himself acquainted with those that are useful. You shall see a man readily ascertain every herb of the field, yet hardly know wheat from barley, or at least one sort of wheat or barley from another.

But of all sorts of vegetation the grasses seem to be most neglected; neither the farmer nor the grazier seem to distinguish the annual from the perennial, the hardy from the tender, nor the succulent and nutritive from the dry and juiceless.

The study of grasses would be of great consequence to a northerly, and grazing kingdom. The botanist that could improve the sward of the district where he lived would be a useful member of society: to raise a thick turf on a naked soil would be worth volumes of systematic knowledge; and he would be the best commonwealth's man that could occasion the growth of 'two blades of grass where one alone was seen before'.*

I am, &c.

Letter 41

Selborne, July 3, 1778.

DEAR SIR,

In a district so diversified with such a variety of hill and dale, aspects, and soils, it is no wonder that great choice of plants should be found. Chalks, clays, sands, sheep-walks and downs, bogs, heaths, woodlands, and champaign fields, cannot but furnish an ample *Flora*. The deep rocky lanes abound with *filices*, and the pastures and moist woods with fungi. If in any branch of botany we may seem to be wanting, it must be in the large aquatic plants, which are not to be expected on a spot far removed from rivers, and lying up amidst the hill country at the spring heads. To enumerate all the plants that have been discovered within our limits would be a needless work; but a short list of the more rare, and the spots where they are to be found, may be neither unacceptable nor unentertaining:*

Helleborus foetidus, stinking hellebore, bear's-foot, or setter-wort—all over the High-wood and Coney-croft-hanger: this continues a great branching plant the winter through, blossoming about January, and is very ornamental in shady walks and shrubberies. The good women give the leaves powdered to children troubled with worms; but it is a violent remedy, and ought to be administered with caution.

Helleborus viridis, green hellebore—in the deep stony lane on the left hand just before the turning to Norton-farm, and at the top of Middle Dorton under the hedge: this plant dies down to the ground early in autumn, and springs again about February, flowering almost as soon as it appears above ground.

Vaccinium oxycoccos, creeping bilberries, or cranberries—in the bogs of Bin's-pond;

Vaccinium myrtillus, whortle, or bilberries—on the dry hillocks of Woolmer-forest;

Drosera rotundifolia, round-leaved sundew. In the bogs
——*longifolia*, long-leaved ditto. of Bin's pond.

Comarum palustre, purple comarum, or marsh cinque foil—in the bogs of Bin's-pond;

Hypericum androsaemum, Tutsan, St John's Wort—in the stony, hollow lanes;

Vinca minor, less periwinkle—in Selborne-hanger and Shrub-wood;

Monotropa hypopitys, yellow monotropa, or bird's nest—in Selborne-hanger under the shady beeches, to whose roots it seems to be parasitical—at the north-west end of the Hanger;

Chlora perfoliata, Blackstonia perfoliata, Hudsoni, perfoliated yellow-wort—on the banks in the King's-field;

Paris quadrifolia, herb Paris, true-love, or one-berry—in the Church-litten-coppice;

Chrysosplenium oppositifolium, opposite golden saxifrage—in the dark and rocky hollow lanes;

Gentiana amarella, autumnal gentian, or felwort—on the Zig-zag and Hanger;

Lathraea squammaria, tooth-wort*—in the Church-litten-coppice under some hazels near the foot-bridge, and in Trimming's garden-hedge, on the dry-wall opposite Grange-yard;

Dipsacus pilosus, small teasel—in the Short and Long Lith;

Lathyrus sylvestris, narrow-leaved, or wild lathyrus—in the bushes at the foot of the Short Lith, near the path;

Ophrys spiralis, ladies' traces—in the Long Lith, and towards the south-corner of the common;

Ophrys nidus avis, birds' nest ophrys—in the Long Lith under the shady beeches among the dead leaves; in Great Dorton among the bushes, and on the Hanger plentifully;

Serapias latifolia, helleborine—in the High-wood under the shady beeches;

Daphne laureola, spurge laurel—in Selborne-Hanger and the High-wood;

Daphne mezereum, the mezereon—in Selborne-Hanger among the shrubs at the south-east end above the cottages;

Lycoperdon tuber, truffles*—in the Hanger and High-wood;

Sambucus ebulus, dwarf elder, wallwort, or danewort—among the rubbish and ruined foundations of the Priory.

Of all the propensities of plants none seem more strange than their different periods of blossoming.* Some produce their flowers in the winter, or very first dawnings of spring; many when the spring is established; some at midsummer, and some not till autumn. When we see the *helleborus foetidus* and *helleborus niger* blowing at Christmas, the *helleborus hyemalis* in January, and the *helleborus viridis* as soon as ever it emerges out of the ground, we do not wonder, because they are kindred plants that we expect should keep pace the one with the other. But other congenerous vegetables differ so widely in their time of flowering, that we cannot but admire. I shall only instance at present in the *crocus sativus*, the vernal, and the autumnal crocus, which have such an affinity, that the best botanists only make them varieties of the same *genus*, of which there is only one species; not being able to discern any difference in the corolla, or in the internal structure. Yet the vernal crocus expands its flowers by the beginning of March at farthest, and often in very rigorous weather; and cannot be retarded but by some violence offered; while the autumnal (the Saffron) defies the influence of the spring and summer, and will not blow till most plants begin to fade and run to seed. This circumstance is one of the wonders of the creation, little noticed, because a common occurrence: yet ought not to be overlooked on account of its being familiar, since it would be as difficult to be explained as the most stupendous phaenomenon in nature.

> Say, what impels, amidst surrounding snow
> Congeal'd, the crocus' flamy bud to glow?

Say, what retards, amidst the summer's blaze,
Th' autumnal bulb, till pale, declining days?
The GOD of SEASONS; whose pervading power
Controls the sun, or sheds the fleecy shower:
He bids each flower his quickening word obey;
Or to each lingering bloom enjoins delay.*

Letter 42

Omnibus animalibus reliquis certus et uniusmodi, et in
suo cuique genere incessus est: aves solae vario meatu
feruntur, et in terra, et in aere.

[All other creatures, in their particular classes, have a distinctive
mode of locomotion; only birds have the capacity to use different
means for moving on land from the means used in the air. Pliny,
Natural History 10. 38]

Selborne, Aug. 7, 1778.

DEAR SIR,

A good ornithologist should be able to distinguish birds by
their air as well as by their colours and shape; on the ground
as well as on the wing, and in the bush as well as in the hand.
For, though it must not be said that every species of birds has
a manner peculiar to itself, yet there is somewhat in most
genera at least, that at first sight discriminates them, and
enables a judicious observer to pronounce upon them with
some certainty. Put a bird in motion

Et vera incessu patuit . . .

[And the kind of bird it is becomes immediately apparent.]*

Thus kites and buzzards sail round in circles with wings
expanded and motionless; and it is from their gliding manner
that the former are still called in the north of England *gleads*,
from the Saxon verb *glidan*, to glide. The kestrel, or windhover, has a peculiar mode of hanging in the air in one place,
his wings all the while being briskly agitated. Hen-harriers fly
low over heaths or fields of corn, and beat the ground regularly

like a pointer or setting-dog. Owls move in a buoyant manner, as if lighter than the air; they seem to want ballast. There is a peculiarity belonging to ravens that must draw the attention even of the most incurious—they spend all their leisure time in striking and cuffing each other on the wing in a kind of playful skirmish; and, when they move from one place to another, frequently turn on their backs with a loud croak, and seem to be falling to the ground. When this odd gesture betides them, they are scratching themselves with one foot, and thus lose the center of gravity. Rooks sometimes dive and tumble in a frolicksome manner; crows and daws swagger in their walk; wood-peckers fly *volatu undoso* [in an undulating fashion], opening and closing their wings at every stroke, and so are always rising or falling in curves. All of this genus use their tails, which incline downward, as a support while they run up trees. Parrots, like all other hooked-clawed birds, walk aukwardly, and make use of their bill as a third foot, climbing and descending with ridiculous caution. All the *gallinae* parade and walk gracefully, and run nimbly; but fly with difficulty, with an impetuous whirring, and in a straight line. Magpies and jays flutter with powerless wings, and make no dispatch; herons seem incumbered with too much sail for their light bodies; but these vast hollow wings are necessary in carrying burdens, such as large fishes, and the like; pigeons, and particularly the sort called smiters, have a way of clashing their wings the one against the other over their backs with a loud snap; another variety called tumblers turn themselves over in the air. Some birds have movements peculiar to the season of love: thus ring-doves, though strong and rapid at other times, yet in the spring hang about on the wing in a toying and playful manner; thus the cock-snipe, while breeding, forgetting his former flight, fans the air like the windhover; and the green-finch in particular exhibits such languishing and faultering gestures as to appear like a wounded and dying bird; the king-fisher darts along like an arrow; fern-owls, or goat-suckers, glance in the dusk over the tops of trees like a meteor; starlings as it were swim along, while missel-thrushes use a wild and desultory flight; swallows sweep over the surface of the ground and water, and distinguish themselves by rapid turns and quick evolutions; swifts dash round in circles; and the bank-martin moves with frequent vacillations like a butter-fly. Most of the small birds fly by jerks, rising and falling as they advance. Most small birds hop; but wagtails and larks walk, moving their legs alternately. Skylarks rise and fall perpendicularly as they sing; woodlarks hang poised in the air; and titlarks rise and fall in large curves, singing in their descent. The white-throat uses odd jerks and gesticulations over the tops of hedges and bushes. All the duck-kind waddle; divers and auks walk as if fettered, and stand erect on their tails: these are the *compedes* of Linnaeus. Geese and cranes, and most wild-fowls, move in figured flights, often changing their position. The secondary remiges of *Tringae*, wild-ducks, and some others, are very long, and give their wings, when in motion, an hooked appearance. Dabchicks, moor-hens, and coots, fly erect, with their legs hanging down, and hardly make any dispatch: the reason is plain, their wings are placed too forward out of the true center of gravity; as the legs of auks and divers are situated too backward.

Letter 43

Selborne, Sept. 9, 1778.

DEAR SIR,

From the motion of birds, the transition is natural enough to their notes and language, of which I shall say something. Not that I would pretend to understand their language like the vizier; who, by the recital of a conversation which passed between two owls, reclaimed a sultan,[23] before delighting in conquest and devastation; but I would be thought only to mean that many of the winged tribes have various sounds and voices adapted to express their various passions, wants, and feelings; such as anger, fear, love, hatred, hunger, and the like. All species are not equally eloquent; some are copious and fluent as it were in their utterance, while others are confined to a few important sounds: no bird, like the fish kind, is quite mute, though some are rather silent. The language of birds is very ancient, and, like other ancient modes of speech, very elliptical; little is said, but much is meant and understood.*

The notes of the eagle-kind are shrill and piercing; and about the season of nidification much diversified, as I have been often assured by a curious observer of Nature, who long resided at Gibraltar, where eagles abound. The notes of our hawks much resemble those of the king of birds. Owls have very expressive notes; they hoot in a fine vocal sound, much resembling the *vox humana* [human voice], and reducible by a pitch-pipe to a musical key. This note seems to express complacency and rivalry among the males: they use also a quick call and an horrible scream; and can snore and hiss when they mean to menace. Ravens, besides their loud croak, can exert a deep and solemn note that makes the woods to echo; the amorous sound of a crow is strange and ridiculous; rooks, in the breeding season, attempt sometimes in the gaiety of their hearts to sing, but with no great success; the parrot-kind have many modulations of voice, as appears by their aptitude to learn human sounds; doves coo in an amorous and mournful manner, and are emblems of despairing lovers; the woodpecker sets up a sort of loud and hearty laugh; the fern-owl, or goat-sucker, from the dusk till day-break, serenades his mate with the clattering of castanets. All the tuneful *passeres* express their complacency by sweet modulations, and a variety of melody. The swallow, as has been observed in a former letter, by a shrill alarm bespeaks the attention of the other hirundines, and bids them be aware that the hawk is at hand. Aquatic and gregarious birds, especially the nocturnal, that shift their quarters in the dark, are very noisy and loquacious; as cranes, wild-geese, wild-ducks, and the like: their perpetual clamour prevents them from dispersing and losing their companions.

In so extensive a subject, sketches and outlines are as much as can be expected: for it would be endless to instance in all the infinite variety of the feathered nation. We shall therefore confine the remainder of this letter to the few domestic fowls of our yards, which are most known, and therefore best understood. And first the peacock, with his gorgeous train, demands our attention; but, like most of the gaudy birds, his notes are grating and shocking to the ear: the yelling of cats, and the braying of an ass, are not more disgustful. The voice of the goose is trumpet-like, and clanking; and once saved the Capitol at Rome, as grave historians assert:* the hiss also of the gander is formidable and full of menace, and 'protective of his young.' Among ducks the sexual distinction of voice is remarkable; for, while the quack of the female is loud and sonorous, the voice of the drake is inward and harsh, and feeble, and scarce discernible. The cock turkey struts and gobbles to his mistress in a most uncouth manner; he hath also a pert and petulant note when he attacks his adversary. When a hen turkey leads forth her young brood she keeps a watchful eye; and if a bird of prey appear, though ever so high in the air, the careful mother announces the enemy with a little inward moan, and watches him with a steady and attentive look; but, if he approach, her note becomes earnest and alarming, and her outcries are redoubled.

No inhabitants of a yard seem possessed of such a variety of expression and so copious a language as common poultry. Take a chicken of four or five days old, and hold it up to a window where there are flies, and it will immediately seize its prey, with little twitterings of complacency; but if you tender it a wasp or a bee, at once its note becomes harsh, and expressive of disapprobation and a sense of danger. When a pullet is ready to lay she intimates the event by a joyous and easy soft note. Of all the occurrences of their life that of laying seems to be the most important; for no sooner has a hen disburdened herself, than she rushes forth with a clamorous kind of joy, which the cock and the rest of his mistresses immediately adopt. The tumult is not confined to the family concerned, but catches from yard to yard, and spreads to every homestead within hearing, till at last the whole village is in an uproar. As soon as a hen becomes a mother her new relation demands a new language; she then runs clocking and scream-ing about, and seems agitated as if possessed. The father of the flock has also a considerable vocabulary; if he finds food, he calls a favourite concubine to partake; and if a bird of prey passes over, with a warning voice he bids his family beware. The gallant chanticleer has, at command, his amorous phrases and his terms of defiance. But the sound by which he is best known is his crowing: by this he has been distinguished in all ages as the countryman's clock or larum, as the watchman that proclaims the divisions of the night. Thus the poet elegantly styles him:

> . . . the crested cock, whose clarion sounds
> The silent hours.

> [Milton, *Paradise Lost*, VII]

A neighbouring gentleman one summer had lost most of his chickens by a sparrow-hawk, that came gliding down between a faggot pile and the end of his house to the place where the coops stood. The owner, inwardly vexed to see his flock thus diminishing, hung a setting net adroitly between the pile and the house, into which the caitiff dashed, and was entangled. Resentment suggested the law of retaliation; he therefore clipped the hawk's wings, cut off his talons, and, fixing a cork on his bill, threw him down among the brood-hens. Imagination cannot paint the scene that ensued; the expressions that fear, rage, and revenge, inspired, were new, or at least such as had been unnoticed before: the exasperated matrons upbraided, they execrated, they insulted, they triumphed. In a word, they never desisted from buffeting their adversary till they had torn him in an hundred pieces.

Letter 44

Selborne.

. . . Monstrent
. . . .
Quid tantum Oceano properent se tingere soles
Hyberni; vel quae tardis mora noctibus obstet.

[Let (the Muses) show . . . why the winter sun
    hastens so fast (across the sky) only to dip
    into the ocean; or, what impediments hinder
    the slow-moving nights.   Virgil, *Georgics* 2]

GENTLEMEN who have outlets* might contrive to make ornament subservient to utility: a pleasing eye-trap might also contribute to promote science: an obelisk in a garden or park might be both an embellishment and an heliotrope.*

Any person that is curious, and enjoys the advantage of a good horizon, might, with little trouble, make two heliotropes; the one for the winter, the other for the summer solstice: and these two erections might be constructed with very little expense; for two pieces of timber frame-work, about ten or twelve feet high, and four feet broad at the base, and close lined with plank, would answer the purpose.

The erection for the former should, if possible, be placed within sight of some window in the common sitting parlour; because men, at that dead season of the year, are usually within doors at the close of the day; while that for the latter might be fixed for any given spot in the garden or outlet: whence the owner might contemplate, in a fine summer's evening, the utmost extent that the sun makes to the north-ward at the season of the longest days. Now nothing would be necessary but to place these two objects with so much exactness, that the westerly limb of the sun, at setting, might but just clear the winter heliotrope to the west of it on the shortest day; and that the whole disc of the sun, at the longest day, might exactly at setting also clear the summer heliotrope to the north of it.

By this simple expedient it would soon appear that there is no such thing, strictly speaking, as a solstice; for, from the shortest day, the owner would, every clear evening, see the disc advancing, at its setting, to the westward of the object; and, from the longest day, observe the sun retiring backwards every evening at its setting, towards the object westward, till, in a few nights, it would set quite behind it, and so by degrees to the west of it: for when the sun comes near the summer solstice, the whole disc of it would at first set behind the object; after a time the northern limb would first appear, and so every night gradually more, till at length the whole diameter would set northward of it for about three nights; but on the middle night of the three, sensibly more remote than the former or following. When beginning its recess from the summer tropic,

it would continue more and more to be hidden every night, till
at length it would descend quite behind the object again; and
so nightly more and more to the westward.

Letter 45

Selborne.

. . . Mugire videbis
Sub pedibus terram, et descendere montibus ornos

[You will experience the earth rumbling beneath
    your feet and the ash trees falling down the
    mountainsides.   Virgil, *Aeneid* 4]

WHEN I was a boy I used to read, with astonishment and
implicit assent, accounts in *Baker's Chronicle* of walking hills
and travelling mountains. John Philips, in his *Cyder*, alludes to
the credit that was given to such stories with a delicate but
quaint vein of humour peculiar to the author of the *Splendid
Shilling*.

> I nor advise, nor reprehend the choice
> Of Marcley Hill; the apple no where finds
> A kinder mould; yet 'tis unsafe to trust
> Deceitful ground: who knows but that once more
> This mount may journey, and his present site
> Forsaken, to thy neighbour's bounds transfer
> Thy goodly plants, affording matter strange
> For law debates!*

But, when I came to consider better, I began to suspect that
though our hills may never have journeyed far, yet that the
ends of many of them have slipped and fallen away at distant
periods, leaving the cliffs bare and abrupt. This seems to have
been the case with Nore and Whetham Hills; and especially
with the ridge between Harteley Park and Ward le ham, where
the ground has slid into vast swellings and furrows; and lies
still in such romantic confusion as cannot be accounted for

from any other cause. A strange event, that happened not long since, justifies our suspicions; which, though it befell not within the limits of this parish, yet as it was within the hundred of Selborne,* and as the circumstances were singular, may fairly claim a place in a work of this nature.

The months of January and February, in the year 1774, were remarkable for great melting snows and vast gluts of rain; so that by the end of the latter month the land-springs, or lavants, began to prevail, and to be near as high as in the memorable winter of 1764. The beginning of March also went on in the same tenor, when, in the night between the 8th and 9th of that month, a considerable part of the great woody hanger at Hawkley was torn from its place, and fell down, leaving a high freestone cliff naked and bare, and resembling the steep side of a chalk-pit. It appears that this huge fragment, being perhaps sapped and undermined by waters, foundered, and was ingulfed, going down in a perpendicular direction; for a gate which stood in the field, on the top of the hill, after sinking with its posts for thirty or forty feet, remained in so true and upright a position as to open and shut with great exactness, just as in its first situation. Several oaks are also still standing, and in a state of vegetation, after taking the same desperate leap.* That great part of this prodigious mass was absorbed in some gulf below, is plain also from the inclining ground at the bottom of the hill, which is free and unincumbered; but would have been buried in heaps of rubbish, had the fragment parted and fallen forward. About an hundred yards from the foot of this hanging coppice stood a cottage by the side of a lane; and two hundred yards lower, on the other side of the lane, was a farm-house, in which lived a labourer and his family; and, just by, a stout new barn. The cottage was inhabited by an old woman and her son, and his wife. These people, in the evening, which was very dark and tempestuous, observed that the brick floors of their kitchens began to heave and part; and that the walls seemed to open, and the roofs to crack: but they all agree that no tremor of the ground, indicating an earthquake, was ever felt; only that the wind continued to make a most tremendous roaring in the woods and hangers. The miserable inhabitants, not daring to go to bed, remained in the utmost solicitude and confusion, expecting every moment to be buried under the ruins of their shattered edifices. When day-light came they were at leisure to contemplate the devastations of the night: they then found that a deep rift, or chasm, had opened under their houses, and torn them, as it were, in two; and that one end of the barn had suffered in a similar manner; that a pond near the cottage had undergone a strange reverse, becoming deep at the shallow end, and so *vice versa*; that many large oaks were removed out of their perpendicular, some thrown down, and some fallen into the heads of neighbouring trees; and that a gate was thrust forward, with its hedge, full six feet, so as to require a new track to be made to it. From the foot of the cliff the general course of the ground, which is pasture, inclines in a moderate descent for half a mile, and is interspersed with some hillocks, which were rifted, in every direction, as well towards the great woody hanger, as from it. In the first pasture the deep clefts began; and running across the lane, and under the buildings, made such vast shelves that the road was impassable for some time; and so over to an arable field on the other side, which was strangely torn and disordered. The second pasture field, being more soft and springy, was protruded forward without many fissures in the turf, which was raised in long ridges resembling graves, lying at right angles to the motion. At the bottom of this enclosure the soil and turf rose many feet against the bodies of some oaks that obstructed their farther course and terminated this awful commotion.

The perpendicular height of the precipice, in general, is twenty-three yards; the length of the lapse, or slip, as seen from the fields below, one hundred and eighty-one; and a partial fall, concealed in the coppice, extends seventy yards more: so that the total length of this fragment that fell was two hundred and fifty-one yards. About fifty acres of land suffered from this violent convulsion; two houses were entirely destroyed; one end of a new barn was left in ruins, the walls being cracked through the very stones that composed them; a hanging coppice was changed to a naked rock; and some grass grounds and an arable field so broken and rifted by the chasms as to be rendered, for a time, neither fit for the plough or safe for pasturage, till considerable labour and expense had been bestowed in levelling the surface and filling in the gaping fissures.

Letter 46

Selborne.

... *resonant arbusta* ...

[the woods resound (with the shrill voice of the grasshopper). Virgil, *Eclogues* 2]

THERE is a steep abrupt pasture field interspersed with furze close to the back of this village, well known by the name of the Short Lithe, consisting of a rocky dry soil, and inclining to the afternoon sun. This spot abounds with the *gryllus campestris*, or field-cricket; which, though frequent in these parts, is by no means a common insect in many other counties.

As their cheerful summer cry cannot but draw the attention of a naturalist, I have often gone down to examine the economy of these *grylli*, and study their mode of life: but they are so shy and cautious that it is no easy matter to get a sight of them; for, feeling a person's footsteps as he advances, they stop short in the midst of their song, and retire backward nimbly into their burrows, where they lurk till all suspicion of danger is over.

At first we attempted to dig them out with a spade, but without any great success; for either we could not get to the bottom of the hole, which often terminated under a great stone; or else, in breaking up the ground, we inadvertently squeezed the poor insect to death. Out of one so bruised we took a multitude of eggs, which were long and narrow, of a yellow colour, and covered with a very tough skin. By this accident we learned to distinguish the male from the female; the former of which is shining black, with a golden stripe across his shoulders; the latter is more dusky, more capacious about the abdomen, and carries a long sword-shaped weapon at her tail, which probably is the instrument with which she deposits her eggs in crannies and safe receptacles.

Where violent methods will not avail, more gentle means will often succeed; and so it proved in the present case: for, though a spade be too boisterous and rough an implement, a pliant stalk of grass, gently insinuated into the caverns, will probe their windings to the bottom, and quickly bring out the inhabitant; and thus the humane inquirer may gratify his curiosity without injuring the object of it. It is remarkable that, though these insects are furnished with long legs behind, and brawny thighs for leaping, like grasshoppers; yet when driven from their holes they shew no activity, but crawl along in a shiftless manner, so as easily to be taken: and again, though provided with a curious apparatus of wings, yet they never exert them when there seems to be the greatest occasion. The males only make that shrilling noise perhaps out of rivalry and emulation, as is the case with many animals which exert some sprightly note during their breeding-time: it is raised by a brisk friction of one wing against the other. They are solitary beings, living singly male or female, each as it may happen; but there must be a time when the sexes have some intercourse, and then the wings may be useful perhaps during the hours of night. When the males meet they will fight fiercely, as I found by some which I put into the crevices of a dry stone wall, where I should have been glad to have made them settle.* For though they seemed distressed by being taken out of their knowledge, yet the first that got possession of the chinks would seize on any that were obtruded upon them with a vast row of serrated fangs. With their strong jaws, toothed like the shears of a lobster's claws, they perforate and round their curious regular cells, having no fore-claws to dig, like the mole-cricket. When taken in hand I could not but wonder that they never offered to defend themselves, though armed with such formidable weapons. Of such herbs as grow before the mouths of their burrows they eat indiscriminately; and on a little platform, which they make just by, they drop their dung; and never, in the day time, seem to stir more than two or three inches from home. Sitting in the entrance of their caverns they chirp all night as well as day from the middle of the month of

May to the middle of July; and in hot weather, when they are most vigorous, they make the hills echo; and, in the still hours of darkness, may be heard to a considerable distance. In the beginning of the season their notes are more faint and inward; but become louder as the summer advances, and so die away again by degrees.

Sounds do not always give us pleasure according to their sweetness and melody; nor do harsh sounds always displease. We are more apt to be captivated or disgusted with the associations which they promote, than with the notes themselves. Thus the shrilling of the field-cricket, though sharp and stridulous, yet marvellously delights some hearers, filling their minds with a train of summer ideas of every thing that is rural, verdurous, and joyous.

About the tenth of March the crickets appear at the mouths of their cells, which they then open and bore, and shape very elegantly. All that ever I have seen at that season were in their pupa state, and had only the rudiments of wings, lying under a skin or coat,—which must be cast before the insect can arrive at its perfect state;[24] from whence I should suppose that the old ones of last year do not always survive the winter. In August their holes begin to be obliterated, and the insects are seen no more till spring.

Not many summers ago I endeavoured to transplant a colony to the terrace in my garden, by boring deep holes in the sloping turf.* The new inhabitants stayed some time, and fed and sung; but wandered away by degrees, and were heard at a farther distance every morning; so that it appears that on this emergency they made use of their wings in attempting to return to the spot from which they were taken.

One of these crickets, when confined in a paper cage and set in the sun, and supplied with plants moistened with water, will feed and thrive, and become so merry and loud as to be irksome in the same room where a person is sitting: if the plants are not wetted it will die.

Letter 47

Selborne.

Far from all resort of mirth
Save the cricket on the hearth.

(Milton, 'Il Penseroso')

DEAR SIR,

While many other insects must be sought after in fields and woods, and waters, the *gryllus domesticus*, or house-cricket, resides altogether within our dwellings,* intruding itself upon our notice whether we will or no. This species delights in new-built houses, being, like the spider, pleased with the moisture of the walls; and besides, the softness of the mortar enables them to burrow and mine between the joints of the bricks or stones, and to open communications from one room to another. They are particularly fond of kitchens and bakers' ovens, on account of their perpetual warmth.

Tender insects that live abroad either enjoy only the short period of one summer, or else doze away the cold uncomfortable months in profound slumbers; but these, residing as it were in a torrid zone, are always alert and merry: a good Christmas fire is to them like the heats of the dog-days. Though they are frequently heard by day, yet is their natural time of motion only in the night. As soon as it grows dusk, the chirping increases, and they come running forth, and are from the size of a flea to that of their full stature. As one should suppose, from the burning atmosphere which they inhabit, they are a thirsty race, and shew a great propensity for liquids, being found frequently drowned in pans of water, milk, broth, or the like. Whatever is moist they affect; and therefore often gnaw holes in wet woollen stockings and aprons that are hung to the fire: they are the housewife's barometer, foretelling her when it will rain; and are prognostic sometimes, she thinks, of ill or good luck; of the death of a near relation, or the approach of an absent lover. By being the constant companions of her solitary hours they naturally become the objects of her superstition. These crickets are not only very thirsty, but very voracious; for they will eat the scummings of pots, and yeast, salt, and crumbs of bread; and any kitchen offal or sweepings. In the summer we have observed them to fly, when it became dusk, out of the windows, and over the neighbouring roofs. This feat of activity accounts for the sudden manner in which they often leave their haunts, as it does for the method by which they come to houses where they were not known before. It is remarkable, that many sorts of insects seem never to use their wings but when they have a mind to shift their quarters and settle new colonies. When in the air they move *volatu undoso*, in waves or curves, like wood-peckers, opening and shutting their wings at every stroke, and so are always rising or sinking.

When they increase to a great degree, as they did once in the house where I am now writing, they become noisome pests, flying into the candles, and dashing into people's faces; but may be blasted and destroyed by gunpowder discharged into their crevices and crannies. In families, at such times, they are, like Pharaoh's plague of frogs—'in their bedchambers, and upon their beds, and in their ovens, and in their kneading-troughs.'[25] Their shrilling noise is occasioned by a brisk attrition of their wings. Cats catch hearth-crickets, and, playing with them as they do with mice, devour them. Crickets may be destroyed, like wasps, by phials half filled with beer, or any liquid, and set in their haunts; for, being always eager to drink, they will crowd in till the bottles are full.

Letter 48

Selborne.

How diversified are the modes of life not only of incongruous but even of congenerous animals; and yet their specific distinctions are not more various than their propensities. Thus, while the field-cricket delights in sunny dry banks, and the house-cricket rejoices amidst the glowing heat of the kitchen hearth or oven, the *gryllus gryllo talpa* (the mole-cricket), haunts moist meadows, and frequents the sides of ponds and banks of streams, performing all its functions in a swampy wet soil. With a pair of fore-feet, curiously adapted to the purpose, it burrows and works under ground like the mole, raising a ridge as it proceeds, but seldom throwing up hillocks.

As mole-crickets often infest gardens by the sides of canals, they are unwelcome guests to the gardener, raising up ridges in their subterraneous progress, and rendering the walks unsightly. If they take to the kitchen quarters, they occasion great damage among the plants and roots, by destroying whole beds of cabbages, young legumes, and flowers. When dug out they seem very slow and helpless, and make no use of their wings by day; but at night they come abroad, and make long excursions, as I have been convinced by finding stragglers, in a morning, in improbable places. In fine weather, about the middle of April, and just at the close of day, they begin to solace themselves with a low, dull, jarring note, continued for a long time without interruption, and not unlike the chattering of the fern-owl, or goat-sucker, but more inward.

About the beginning of May they lay their eggs, as I was once an eye-witness: for a gardener at an house, where I was on a visit,* happening to be mowing, on the 6th of that month, by the side of a canal, his scythe struck too deep, pared off a large piece of turf, and laid open to view a curious scene of domestic economy:

> . . . ingentem lato dedit ore fenestram:
> Apparet domus intus, et atria longa patescunt:
> Apparent . . . penetralia.
>
> [(Pyrrhus) made a huge, cavernous opening,
>   and the interior of the palace appeared, with
>   its long halls clearly visible; beyond, there
>   were the inner chambers (of Priam).
>   Virgil, *Aeneid* 2]

There were many caverns and winding passages leading to a kind of chamber, neatly smoothed and rounded, and about the size of a moderate snuff-box. Within this secret nursery were deposited near an hundred eggs of a dirty yellow colour, and enveloped in a tough skin, but too lately excluded to contain any rudiments of young, being full of a viscous substance. The eggs lay but shallow, and within the influence of the sun, just under a little heap of fresh-moved mould, like that which is raised by ants.

When mole-crickets fly they move *cursu undoso*, rising and falling in curves, like the other species mentioned before. In different parts of this kingdom people call them fen-crickets, churr-worms, and eve-churrs, all very apposite names.

Anatomists, who have examined the intestines of these insects, astonish me with their accounts; for they say that, from the structure, position, and number of their stomachs, or maws, there seems to be good reason to suppose that this and the two former species ruminate or chew the cud like many quadrupeds!

Letter 49

Selborne, May 7, 1779.

IT is now more than forty years that I have paid some attention to the ornithology of this district, without being able to exhaust the subject: new occurrences still arise as long as any inquiries are kept alive.

In the last week of last month five of those most rare birds, too uncommon to have obtained an English name, but known to naturalists by the terms of *himantopus*, or *loripes*, and *charadrius himantopus*, were shot upon the verge of Frinsham-pond, a large lake belonging to the Bishop of Winchester, and lying between Woolmer-forest and the town of Farnham, in the county of Surrey. The pond keeper says there were three brace in the flock; but that, after he had satisfied his curiosity, he suffered the sixth to remain unmolested. One of these specimens I procured, and found the length of the legs to be so extraordinary, that, at first sight, one might have supposed the shanks had been fastened on to impose on the credulity of the beholder: they were legs in *caricatura*; and had we seen such proportions on a Chinese or Japan screen we should have made large allowances for the fancy of the draughtsman. These birds are of the plover family, and might with propriety be called the stilt plovers. Brisson, under that idea, gives them the apposite name of *l'échasse*. My specimen, when drawn and stuffed with pepper,* weighed only four ounces and a quarter, though the naked part of the thigh measured three inches and an half, and the legs four inches and an half. Hence we may safely assert that these birds exhibit, weight for inches, incomparably the greatest length of legs of any known bird. The flamingo,* for instance, is one of the most long legged birds, and yet it bears no manner of proportion to the *himantopus*; for a cock flamingo weighs, at an average, about four pounds avoirdupois; and his legs and thighs measure usually about twenty inches. But four pounds are fifteen times and a fraction more than four ounces, and one quarter; and if four ounces and a quarter have eight inches of legs, four pounds must have one hundred and twenty inches and a fraction of legs; viz. somewhat more than ten feet; such a monstrous proportion as the world never saw!* If you should try the experiment in still larger birds the disparity would still increase. It must be matter of great curiosity to see the stilt plover move; to observe how it can wield such a length of lever with such feeble muscles as the thighs seem to be furnished with. At best one should expect it to be but a bad walker: but what adds to the wonder is, that it has no back toe. Now without that steady prop to support its steps it must be liable, in speculation, to perpetual vacillations, and seldom able to preserve the true center of gravity.

The old name of *himantopus* is taken from Pliny; and, by an awkward metaphor, implies that the legs are as slender and pliant as if cut out of a thong of leather. Neither Willughby nor Ray, in all their curious researches, either at home or abroad, ever saw this bird. Mr Pennant never met with it in all Great-Britain, but observed it often in the cabinets of the curious at Paris. Hasselquist says that it migrates to Egypt in the autumn: and a most accurate observer of Nature has assured me that he has found it on the banks of the streams in Andalusia.

Our writers record it to have been found only twice in Great-Britain. From all these relations it plainly appears that these long-legged plovers are birds of South Europe, and rarely

Black-winged Stilt

visit our island; and when they do are wanderers and stragglers, and impelled to make so distant and northern an excursion from motives or accidents for which we are not able to account. One thing may be fairly deduced, that these birds come over to us from the continent, since nobody can suppose that a species not noticed once in an age, and of such a remarkable make, can constantly breed unobserved in this kingdom.

Letter 50

Selborne, April 21, 1780.

DEAR SIR,

The old Sussex tortoise, that I have mentioned to you so often, is become my property. I dug it out of its winter dormitory in March last, when it was enough awakened to express its resentments by hissing; and, packing it in a box with earth, carried it eighty miles in post-chaises.* The rattle and hurry of the journey so perfectly roused it that, when I turned it out on a border, it walked twice down to the bottom of my garden; however, in the evening, the weather being cold, it buried itself in the loose mould, and continues still concealed.

As it will be under my eye, I shall now have an opportunity of enlarging my observations on its mode of life, and propensities; and perceive already that, towards the time of coming forth, it opens a breathing-place in the ground near its head, requiring, I conclude, a freer respiration as it becomes more alive. This creature not only goes under the earth from the middle of November to the middle of April, but sleeps great part of the summer; for it goes to bed in the longest days at four in the afternoon, and often does not stir in the morning till late. Besides, it retires to rest for every shower; and does not move at all in wet days.

When one reflects on the state of this strange being, it is a matter of wonder to find that Providence should bestow such a profusion of days, such a seeming waste of longevity, on a reptile that appears to relish it so little as to squander more than two thirds of its existence in a joyless stupor, and be lost to all sensation for months together in the profoundest of slumbers.

While I was writing this letter, a moist and warm afternoon, with the thermometer at 50, brought forth troops of shell-snails; and, at the same juncture, the tortoise heaved up the mould and put out its head; and the next morning came forth, as it were raised from the dead; and walked about till four in the afternoon.* This was a curious coincidence! a very amusing occurrence! to see such a similarity of feelings between the two φερέοικοι! [house-carriers] for so the Greeks called both the shell-snail and the tortoise.

Summer birds are, this cold and backward spring, unusually late: I have seen but one swallow yet. This conformity with the weather convinces me more and more that they sleep in the winter.

*More Particulars respecting the Old Family Tortoise, omitted in the Natural History**

Because we call this creature an abject reptile, we are too apt to undervalue his abilities, and depreciate his powers of instinct. Yet he is, as Mr Pope says of his lord,

> . . . Much too wise to walk into a well;

and has so much discernment as not to fall down an haha; but to stop and withdraw from the brink with the readiest precaution.

Though he loves warm weather he avoids the hot sun; because his thick shell, when once heated, would, as the poet says of solid armour—'scald with safety'. He therefore spends the more sultry hours under the umbrella of a large cabbage-leaf, or amidst the waving forests of an asparagus-bed.

But as he avoids heat in the summer, so, in the decline of the year, he improves the faint autumnal beams, by getting within the reflection of a fruit-wall: and, though he never has read that planes inclining to the horizon receive a greater share of warmth, he inclines his shell, by tilting it against the wall, to collect and admit every feeble ray.

Pitiable seems the condition of this poor embarrassed reptile: to be cased in a suit of ponderous armour, which he cannot lay aside; to be imprisoned, as it were, within his own shell, must preclude, we should suppose, all activity and disposition for enterprize. Yet there is a season of the year (usually the beginning of June) when his exertions are remarkable. He then walks on tip-toe, and is stirring by five in the morning; and, traversing the garden, examines every wicket and interstice in the fences, through which he will escape if possible: and often has eluded the care of the gardener, and wandered to some distant field. The motives that impel him to undertake these rambles seem to be of the amorous kind; his fancy then becomes intent on sexual attachments, which transport him beyond his usual gravity, and induce him to forget for a time his ordinary solemn deportment.

Letter 51

Selborne, Sept. 3, 1781.

I HAVE now read your *Miscellanies* through with much care and satisfaction; and am to return you my best thanks for the honourable mention made in them of me as a naturalist, which I wish I may deserve.*

In some former letters I expressed my suspicions that many of the house-martins do not depart in the winter far from this village. I therefore determined to make some search about the south-east end of the hill, where I imagined they might slumber out the uncomfortable months of winter. But supposing that the examination would be made to the best advantage in the spring, and observing that no martins had appeared by the 11th of April last; on that day I employed some men to explore the shrubs and cavities of the suspected spot. The persons took pains, but without any success; however, a remarkable incident occurred in the midst of our pursuit—while the labourers were at work a house-martin, the first that had been seen this year, came down the village in the sight of several people, and went at once into a nest, where it stayed a short time, and then flew over the houses; for some days after no martins were observed, not till the 16th of April, and then only a pair. Martins in general were remarkably late this year.

Letter 52

Selborne, Sept. 9, 1781.

I HAVE just met with a circumstance respecting swifts, which furnishes an exception to the whole tenor of my observations ever since I have bestowed any attention on that species of hirundines. Our swifts, in general, withdrew this year about the first day of August, all save one pair, which in two or three days was reduced to a single bird. The perseverance of this individual made me suspect that the strongest of motives, that of an attachment to her young, could alone occasion so late a stay. I watched therefore till the twenty-fourth of August, and then discovered that, under the eaves of the church, she attended upon two young, which were fledged, and now put out their white chins from a crevice. These remained till the twenty-seventh, looking more alert every day, and seeming to long to be on the wing. After this day they were missing at once; nor could I ever observe them with their dam coursing round the church in the act of learning to fly, as the first broods evidently do. On the thirty-first I caused the eaves to be searched, but we found in the nest only two callow, dead, stinking swifts, on which a second nest had been formed. This double nest was full of the black shining cases of the *hippoboscae hirundinis*.

The following remarks on this unusual incident are obvious. The first is, that though it may be disagreeble to swifts to remain beyond the beginning of August, yet that they can subsist longer is undeniable. The second is, that this uncommon event, as it was owing to the loss of the first brood, so it corroborates my former remark, that swifts breed regularly but once; since, was the contrary the case, the occurrence above could neither be new nor rare.

P.S.—One swift was seen at Lyndon, in the county of Rutland, in 1782, so late as the third of September.*

Letter 53

As I have sometimes known you make inquiries about several kinds of insects, I shall here send you an account of one sort which I little expected to have found in this kingdom. I had often observed that one particular part of a vine growing on the walls of my house was covered in the autumn with a black dust-like appearance, on which the flies fed eagerly; and that the shoots and leaves thus affected did not thrive; nor did the fruit ripen. To this substance I applied my glasses; but could not discover that it had any thing to do with animal life, as I at first expected: but, upon a closer examination behind the larger boughs, we were suprised to find that they were coated over with husky shells, from whose sides proceeded a cotton-like substance, surrounding a multitude of eggs. This curious and uncommon production put me upon recollecting what I have heard and read, concerning the *coccus vitis viniferae* of Linnaeus, which, in the south of Europe, infests many vines, and is an horrid and loathsome pest.* As soon as I had turned to the accounts given of this insect, I saw at once that it swarmed on my vine; and did not appear to have been at all checked by the preceding winter, which had been uncommonly severe.

Not being then at all aware that it had any thing to do with England, I was much inclined to think that it came from Gibraltar among the many boxes and packages of plants and birds which I had formerly received from thence; and especially as the vine infested grew immediately under my study-window, where I usually kept my specimens. True it is that I had received nothing from thence for some years: but as insects, we know, are conveyed from one country to another in a very unexpected manner, and have a wonderful power of maintaining their existence till they fall into a *nidus* proper for their support and increase, I cannot but suspect still that these *cocci* came to me originally from Andalusia. Yet, all the while, candour obliges me to confess that Mr Lightfoot has written me word that he once, and but once, saw these insects on a vine at Weymouth in Dorsetshire,* which, it is here to be observed, is a sea-port town to which the *coccus* might be conveyed by shipping.

As many of my readers may possibly never have heard of this strange and unusual insect, I shall here transcribe a passage from a natural history of Gibraltar, written by the Reverend John White, late vicar of Blackburn in Lancashire, but not yet published:—*

'In the year 1770 a vine, which grew on the east-side of my house, and which had produced the finest crops of grapes for years past, was suddenly overspread on all the woody branches with large lumps of a white fibrous substance resembling spiders' webs, or rather raw cotton. It was of a very clammy quality, sticking fast to every thing that touched it, and capable of being spun into long threads. At first I suspected it to be the product of spiders, but could find none. Nothing was to be seen connected with it but many brown oval husky shells, which by no means looked like insects, but rather resembled bits of the dry bark of the vine. The tree had a plentiful crop of grapes set, when this pest appeared upon it; but the fruit was manifestly injured by this foul incumbrance. It remained all the summer, still increasing, and loaded the woody and bearing branches to a vast degree. I often pulled off great quantities by handfuls; but it was so slimy and tenacious that it could by no means be cleared. The grapes never filled to their natural perfection, but turned watery and vapid. Upon perusing the works afterwards of M. de Réaumur, I found this matter perfectly described and accounted for. Those husky shells, which I had observed, were no other than the female *coccus*, from whose sides this cotton-like substance exudes, and serves as a covering and security for their eggs.'

To this account I think proper to add, that, though the female *cocci* are stationary, and seldom remove from the place to which they stick, yet the male is a winged insect; and that the black dust which I saw was undoubtedly the excrement of the females, which is eaten by ants as well as flies. Though the utmost severity of our winter did not destroy these insects, yet the attention of the gardener in a summer or two has entirely relieved my vine from this filthy annoyance.*

As we have remarked above that insects are often conveyed from one country to another in a very unaccountable manner, I shall here mention an emigration of small *aphides*, which was observed in the village of Selborne no longer ago than August the 1st, 1785.

At about three o'clock in the afternoon of that day, which was very hot, the people of this village were surprised by a shower of *aphides*, or smother-flies, which fell in these parts. Those that were walking in the street at that juncture found themselves covered with these insects, which settled also on the hedges and gardens, blackening all the vegetables where they alighted. My annuals were discoloured with them, and the stalks of a bed of onions were quite coated over for six days after. These armies were then, no doubt, in a state of emigration, and shifting their quarters; and might have come, as we know, from the great hop-plantations of Kent or Sussex, the wind being all that day in the easterly quarter. They were observed at the same time in great clouds about Farnham and all along the vale from Farnham to Alton.[26]

Letter 54

DEAR SIR,

When I happen to visit a family where gold and silver fishes are kept in a glass bowl, I am always pleased with the occurrence, because it offers me an opportunity of observing the actions and propensities of those beings with whom we can be little acquainted in their natural state. Not long since I spent a fortnight at the house of a friend where there was such a vivary, to which I paid no small attention, taking every occasion to remark what passed within its narrow limits.* It was here that I first observed the manner in which fishes die. As soon as the creature sickens, the head sinks lower and lower, and it stands as it were on its head; till, getting weaker, and losing all poise, the tail turns over, and at last it floats on the surface of the water with its belly uppermost. The reason why fishes, when dead, swim in that manner is very obvious; because, when the body is no longer balanced by the fins of the belly, the broad muscular back preponderates by its own gravity, and turns the belly uppermost, as lighter from its being a cavity, and because it contains the swimming-bladders, which contribute to render it buoyant. Some that delight in gold and silver fishes have adopted a notion that they need no aliment. True it is that they will subsist for a long time without any apparent food but what they can collect from pure water frequently changed; yet they must draw some support from animalcula, and other nourishment supplied by the water; because, though they seem to eat nothing, yet the consequences of eating often drop from them. That they are best pleased with such jejune diet may easily be confuted, since if you toss them crumbs they will seize them with great readiness, not to say greediness: however, bread should be given sparingly, lest, turning sour, it corrupt the water. They will also feed on the water-plant called *lemna* (duck's meat), and also on small fry.

When they want to move a little they gently protrude themselves with their *pinnae pectorales*; but it is with their strong muscular tails only that they and all fishes shoot along with such inconceivable rapidity. It has been said that the eyes of fishes are immoveable: but these apparently turn them forward or backward in their sockets as their occasions require. They take little notice of a lighted candle, though applied close to their heads, but flounce and seem much frightened by a sudden stroke of the hand against the support whereon the bowl is hung; especially when they have been motionless, and are perhaps asleep. As fishes have no eyelids, it is not easy to discern when they are sleeping or not, because their eyes are always open.

Nothing can be more amusing than a glass bowl containing such fishes: the double refractions of the glass and water represent them, when moving, in a shifting and changeable variety of dimensions, shades, and colours; while the two mediums, assisted by the concavo-convex shape of the vessel, magnify and distort them vastly; not to mention that the introduction of another element and its inhabitants into our parlours engages the fancy in a very agreeable manner.

Gold and silver fishes, though originally natives of China and Japan, yet are become so well reconciled to our climate as to thrive and multiply very fast in our ponds and stews.* Linnaeus ranks this species of fish under the genus of *cyprinus*, or carp, and calls it *cyprinus auratus*.

Some people exhibit this sort of fish in a very fanciful way; for they cause a glass bowl to be blown with a large hollow space within, that does not communicate with it. In this cavity they put a bird occasionally; so that you may see a goldfinch or a linnet hopping as it were in the midst of the water, and the fishes swimming in a circle round it. The simple exhibition of the fishes is agreeable and pleasant; but in so complicated a way becomes whimsical and unnatural, and liable to the objection due to him,

Qui variare cupit rem prodigialiter unam

[who delights to speculate on a subject
that will scarcely bear it.

Horace, *Ars Poetica*]

I am, &c.

Letter 55

Oct. 10, 1781.

DEAR SIR,

I think I have observed before that much the most considerable part of the house-martins withdraw from hence about the
first week in October; but that some, the latter broods I am
now convinced, linger on till towards the middle of that month:
and that at times, once perhaps in two or three years, a flight,
for one day only, has shown itself in the first week in
November.*

Having taken notice, in October 1780, that the last flight
was numerous, amounting perhaps to one hundred and fifty;
and that the season was soft and still; I was resolved to pay
uncommon attention to these late birds; to find, if possible,
where they roosted, and to determine the precise times of their
retreat. The mode of life of these latter hirundines is very
favourable to such a design; for they spend the whole day in
the sheltered district, between me and the Hanger, sailing
about in a placid, easy manner, and feasting on those insects
which love to haunt a spot so secure from ruffling winds. As
my principal object was to discover the place of their roosting,
I took care to wait on them before they retired to rest, and was
much pleased to find that, for several evenings together, just
at a quarter past five in the afternoon, they all scudded away
in great haste towards the south-east, and darted down among
the low shrubs above the cottages at the end of the hill.* This
spot in many respects seems to be well calculated for their
winter residence: for in many parts it is as steep as the roof of
any house, and therefore secure from the annoyances of water;

and it is moreover clothed with beechen shrubs, which, being stunted and bitten by sheep, make the thickest covert imaginable; and are so entangled as to be impervious to the smallest spaniel: besides, it is the nature of underwood beech never to cast its leaf all the winter; so that, with the leaves on the ground and those on the twigs, no shelter can be more complete. I watched them on to the thirteenth and fourteenth of October, and found their evening retreat was exact and uniform; but after this they made no regular appearance. Now and then a straggler was seen; and, on the twenty-second of October, I observed two in the morning over the village, and with them my remarks for the season ended.

From all these circumstances put together, it is more than probable that this lingering flight, at so late a season of the year, never departed from the island. Had they indulged me that autumn with a November visit, as I much desired, I presume that, with proper assistants, I should have settled the matter past all doubt; but though the third of November was a sweet day, and in appearance exactly suited to my wishes, yet not a martin was to be seen; and so I was forced, reluctantly, to give up the pursuit.

I have only to add that were the bushes, which cover some acres, and are not my own property, to be grubbed and carefully examined, probably those late broods, and perhaps the whole aggregate body of the house-martins of this district, might be found there, in different secret dormitories; and that, so far from withdrawintg into warmer climes, it would appear that they never depart three hundred yards from the village.

Letter 56

THEY who write on natural history cannot too frequently advert to instinct, that wonderful limited faculty, which, in some instances, raises the brute creation as it were above reason, and in others leaves them so far below it. Philosophers have defined instinct to be that secret influence by which every species is impelled naturally to pursue, at all times, the same way or track, without any teaching or example; whereas reason, without instruction, would often vary and do that by many methods which instinct effects by one alone. Now this maxim must be taken in a qualified sense; for there are instances in which instinct does vary and conform to the circumstances of place and convenience.

It has been remarked that every species of bird has a mode of nidification peculiar to itself; so that a school-boy would at once pronounce on the sort of nest before him. This is the case among fields and woods, and wilds; but, in the villages round London, where mosses and gossamer, and cotton from vegetables, are hardly to be found, the nest of the chaffinch has not that elegant finished appearance, nor is it so beautifully studded with lichens, as in a more rural district: and the wren is obliged to construct its house with straws and dry grasses, which do not give it that rotundity and compactness so remarkable in the edifices of that little architect. Again, the regular nest of the house-martin is hemispheric; but where a rafter, or a joist, or a cornice, may happen to stand in the way, the nest is so contrived as to conform to the obstruction, and becomes flat or oval, or compressed.

In the following instances instinct is perfectly uniform and consistent. There are three creatures, the squirrel, the field-mouse, and the bird called the nut-hatch (*sitta Europaea*), which live much on hazle-nuts; and yet they open them each in a different way. The first, after rasping off the small end, splits the shell into two with his long fore-teeth, as a man does with his knife; the second nibbles a hole with his teeth, so regular as if drilled with a wimble, and yet so small that one would wonder how the kernel can be extracted through it; while the last picks an irregular ragged hole with its bill: but as this artist has no paws to hold the nut firm while he pierces it, like an adroit workman, he fixes it, as it were in a vice, in some cleft of a tree, or in some crevice; when, standing over it, he perforates the stubborn shell. We have often placed nuts in the chink of a gate-post where nut-hatches have been known to haunt, and have always found that those birds have readily penetrated them. While at work they make a rapping noise that may be heard at a considerable distance.*

You that understand both the theory and practical part of music may best inform us why harmony or melody should so strangely affect some men, as it were by recollection, for days after a concert is over.* What I mean the following passage will most readily explain:

Praehabebat porro vocibus humanis, instrumentisque harmonicis musicam illam avium: non quod alia quoque non delectaretur; sed quod ex musica humana relinqueretur in animo continens quaedam, attentionemque et somnum conturbans agitatio: dum ascensus, exscensus, tenores, ac mutationes illae sonorum, et consonantiarum euntque, redeuntque per phantasiam:—cum nihil tale relinqui possit ex modulationibus avium, quae, quod non sunt perinde a nobis imitabiles, non possunt perinde internam facultatem commovere.

[He preferred the music of birds to that made either by the human voice or by instruments. It was not that these last failed to give him pleasure; on the contrary, the melodies and harmony would run so repeatedly in his imagination that the continual pleasure distracted his mind and disturbed his sleep. Such, however, was not the case with the music of birds—for the very reason that these melodies do not lend themselves to be repeated by men and, hence, do not stay in the mind to vex and confuse the inner faculties.] Gassendus in *Vita Peireskii.**

This curious quotation strikes me much by so well representing my own case, and by describing what I have so often felt, but never could so well express. When I hear fine music I am haunted with passages therefrom night and day; and especially at first waking, which, by their importunity, give me more uneasiness than pleasure: elegant lessons still tease my imagination, and recur irresistibly to my recollection at seasons, and even when I am desirous of thinking of more serious matters.

I am, &c.

Letter 57

A RARE, and I think a new, little bird frequents my garden, which I have great reason to think is the petti-chaps: it is common in some parts of the kingdom; and I have received formerly several dead specimens from Gibraltar. This bird much resembles the white-throat, but has a more white or rather silvery breast and belly; is restless and active, like the willow-wrens, and hops from bough to bough, examining every part for food; it also runs up the stems of the crown-imperials, and, putting its head into the bells of those flowers, sips the liquor which stands in the *nectarium* of each petal. Sometimes it feeds on the ground like the hedge-sparrow, by hopping about on the grass-plots and mown walks.*

One of my neighbours, an intelligent and observing man, informs me that, in the beginning of May, and about ten minutes before eight o'clock in the evening, he discovered a great cluster of house-swallows, thirty at least he supposes, perching on a willow that hung over the verge of James Knight's upper-pond. His attention was first drawn by the twittering of these birds, which sat motionless in a row on the bough, with their heads all one way, and, by their weight, pressing down the twig so that it nearly touched the water. In this situation he watched them till he could see no longer. Repeated accounts of this sort, spring and fall, induce us greatly to suspect that house-swallows have some strong attachment to water, independent of the matter of food; and, though they may not retire into that element, yet they may conceal themselves in the banks of pools and rivers during the uncomfortable months of winter.

One of the keepers of Wolmer-forest sent me a peregrine-falcon, which he shot on the verge of that district as it was devouring a wood-pigeon.* The *falco peregrinus*, or haggard falcon, is a noble species of hawk seldom seen in the southern counties. In winter 1767 one was killed in the neighbouring parish of Faringdon, and sent by me to Mr Pennant into North-Wales.[27] Since that time I have met with none till now. The specimen mentioned above was in fine preservation, and not injured by the shot: it measured forty-two inches from wing to wing, and twenty-one from beak to tail, and weighed two pounds and an half standing weight. This species is very robust, and wonderfully formed for rapine: its breast was plump and muscular; its thighs long, thick, and brawny; and its legs remarkably short and well set: the feet were armed

Peregrine Falcon

with most formidable, sharp, long talons: the eyelids and cere
of the bill were yellow; but the irides of the eyes dusky; the
beak was thick and hooked, and of a dark colour, and had a
jagged process near the end of the upper mandible on each
side: its tail, or train, was short in proportion to the bulk of its
body: yet the wings, when closed, did not extend to the end of
the train. From its large and fair proportions it might be
supposed to have been a female; but I was not permitted to
cut open the specimen. For one of the birds of prey, which are
usually lean, this was in high case: in its craw were many
barleycorns, which probably came from the crop of the wood-
pigeon, on which it was feeding when shot: for voracious birds
do not eat grain; but, when devouring their quarry, with
undistinguishing vehemence swallow bones and feathers, and
all matters, indiscriminately. This falcon was probably driven

from the mountains of North Wales or Scotland, where they are known to breed, by rigorous weather and deep snows that had lately fallen.

                                        I am, &c.

Letter 58

M Y near neighbour, a young gentleman* in the service of the East-India Company, has brought home a dog and a bitch of the Chinese breed from Canton; such as are fattened in that country for the purpose of being eaten: they are about the size of a moderate spaniel; of a pale yellow colour, with coarse bristling hairs on their backs; sharp upright ears, and peaked heads, which give them a very fox-like appearance. Their hind legs are unusually straight, without any bend at the hock or ham, to such a degree as to give them an awkward gait when they trot. When they are in motion their tails are curved high over their backs like those of some hounds, and have a bare place each on the outside from the tip midway, that does not seem to be matter of accident, but somewhat singular. Their eyes are jet-black, small, and piercing; the insides of their lips and mouths of the same colour, and their tongues blue. The bitch has a dew-claw on each hind leg; the dog has none. When taken out into a field the bitch showed some disposition for hunting, and dwelt on the scent of a covey of partridges till she sprung them, giving her tongue all the time. The dogs in South America are dumb; but these bark much in a short thick manner, like foxes; and have a surly, savage demeanour like their ancestors, which are not domesticated, but bred up in sties, where they are fed for the table with rice-meal and other farinaceous food. These dogs, having been taken on board as soon as weaned, could not learn much from their dam; yet they did not relish flesh when they came to England. In the islands of the Pacific ocean the dogs are bred up on vegetables, and would not eat flesh when offered them by our circumnavigators.

We believe that all dogs, in a state of nature, have sharp,

upright fox-like ears; and that hanging ears, which are esteemed so graceful, are the effect of choice breeding and cultivation. Thus, in the *Travels* of Ysbrandt Ides from Muscovy to China, the dogs which draw the Tartars on snowsledges near the river Oby are engraved with prick-ears, like those from Canton. The Kamschatdales also train the same sort of sharp-eared peak-nosed dogs to draw their sledges; as may be seen in an elegant print engraved for Captain Cook's last voyage round the world.

Now we are upon the subject of dogs, it may not be impertinent to add, that spaniels, as all sportsmen know, though they hunt partridges and pheasants as it were by instinct, and with much delight and alacrity, yet will hardly touch their bones when offered as food; nor will a mongrel dog of my own, though he is remarkable for finding that sort of game. But, when we came to offer the bones of partridges to the two Chinese dogs, they devoured them with much greediness, and licked the platter clean.

No sporting dogs will flush woodcocks till inured to the scent and trained to the sport, which they then pursue with vehemence and transport; but then they will not touch their bones, but turn from them with abhorrence, even when they are hungry.

Now, that dogs should not be fond of the bones of such birds as they are not disposed to hunt is no wonder; but why they reject and do not care to eat their natural game is not so easily accounted for, since the end of hunting seems to be, that the chase pursued should be eaten. Dogs again will not devour the more rancid water-fowls, nor indeed the bones of any wildfowls; nor will they touch the foetid bodies of birds that feed on offal and garbage: and indeed there may be somewhat of providential instinct in this circumstance of dislike; for vultures,[28] and kites, and ravens, and crows, &c. were intended to be messmates with dogs[29] over their carrion; and seem to be appointed by Nature as fellow-scavengers to remove all cadaverous nuisances from the face of the earth.

I am, &c.

Letter 59

THE fossil wood buried in the bogs of Wolmer-forest is not yet all exhausted; for the peat-cutters now and then stumble upon a log. I have just seen a piece which was sent by a labourer of Oakhanger to a carpenter of this village; this was the butt-end of a small oak, about five feet long, and about five inches in diameter. It had apparently been severed from the ground by an axe, was very ponderous, and as black as ebony. Upon asking the carpenter for what purpose he had procured it; he told me that it was to be sent to his brother, a joiner at Farnham, who was to make use of it in cabinet work, by inlaying it along with whiter woods.

Those that are much abroad on evenings after it is dark, in spring and summer, frequently hear a nocturnal bird passing by on the wing, and repeating often a short quick note. This bird I have remarked myself, but never could make out till lately. I am assured now that it is the Stone-curlew (*charadrius oedicnemus*). Some of them pass over or near my house almost every evening after it is dark,* from the uplands of the hill and North field, away down towards Dorton; where, among the streams and meadows, they find a greater plenty of food. Birds that fly by night are obliged to be noisy; their notes often repeated become signals or watchwords to keep them together, that they may not stray or lose each other in the dark.

The evening proceedings and manoeuvres of the rooks are curious and amusing in the autumn. Just before dusk they return in long strings from the foraging of the day, and rendezvous by thousands over Selborne-down, where they wheel round in the air, and sport and dive in a playful manner, all the while exerting their voices, and making a loud cawing, which, being blended and softened by the distance that we at the village are below them, becomes a confused noise or chiding; or rather a pleasing murmur, very engaging to the imagination, and not unlike the cry of a pack of hounds in hollow, echoing woods, or the rushing of the wind in tall trees, or the tumbling of the tide upon a pebbly shore. When this ceremony is over, with the last gleam of day, they retire for the night to the deep beechen woods of Tisted and Ropley. We remember a little girl who, as she was going to bed, used to remark on such an occurrence, in the true spirit of physico-theology, that the rooks were saying their prayers; and yet this child was much too young to be aware that the scriptures have said of the Deity—that 'he feedeth the ravens who call upon him'.*

I am, &c.

Letter 60

In reading Dr Huxham's *Observationes de Aere*, &c. written at Plymouth, I find by those curious and accurate remarks,* which contain an account of the weather from the year 1727 to the year 1748 inclusive, that though there is frequent rain in that district of Devonshire, yet the quantity falling is not great; and that some years it has been very small: for in 1731 the rain measured only 17$^{inch.}$—266$^{thou.}$, and in 1741, 20—354; and again, in 1743, only 20—908. Places near the sea have frequent scuds, that keep the atmosphere moist, yet do not reach far up into the country; making thus the maritime situations appear wet, when the rain is not considerable. In the wettest years at Plymouth the Doctor measured only once 36; and again once, viz. 1734, 37—114: a quantity of rain that has twice been exceeded at Selborne in the short period of my observations. Dr Huxham remarks that frequent small rains keep the air moist; while heavy ones render it more dry, by beating down the vapours. He is also of opinion that the dingy, smoky appearance in the sky, in very dry seasons, arises from the want of moisture sufficient to let the light through, and render the atmosphere transparent; because he had observed several bodies more diaphanous when wet than dry; and did never recollect that the air had that look in rainy seasons.

My friend, who lives just beyond the top of the down, brought his three swivel guns to try them in my outlet, with their muzzles towards the Hanger, supposing that the report would have had a great effect; but the experiment did not answer his expectation. He then removed them to the Alcove on the Hanger; when the sound, rushing along the Lythe and Comb-wood, was very grand: but it was at the Hermitage that the echoes and repercussions delighted the hearers; not only filling the Lythe with the roar, as if all the beeches were tearing up by the roots; but, turning to the left, they pervaded the vale above Combwood-ponds; and after a pause seemed to take up the crash again, and to extend round Harteley-hangers, and to die away at last among the coppices and coverts of Ward le ham. It has been remarked before that this district is an anathoth, a place of responses or echoes, and therefore proper for such experiments: we may farther add that the pauses in echoes, when they cease, and yet are taken up again, like the pauses in music, surprise the hearers, and have a fine effect on the imagination.

The gentleman abovementioned has just fixed a barometer in his parlour at Newton Valence.* The tube was first filled here (at Selborne) twice with care, when the mercury agreed and stood exactly with my own; but, being filled again twice at Newton, the mercury stood, on account of the great elevation of that house, three-tenths of an inch lower than the barometers at this village, and so continues to do, be the weight of the atmosphere what it may. The plate of the barometer at Newton is figured as low as 27; because in stormy weather the mercury there will sometimes descend below 28. We have supposed Newton-house to stand two hundred feet higher than this house: but if the rule holds good, which says that mercury in a barometer sinks one-tenth of an inch for every hundred feet elevation, then the Newton barometer, by standing three-tenths lower than that of Selborne, proves that Newton-house must be three hundred feet higher than that in which I am writing, instead of two hundred.

It may not be impertinent to add, that the barometer at Selborne stand three-tenths of an inch lower than the barometers at South Lambeth: whence we may conclude that the former place is about three hundred feet higher than the latter; and with good reason, because the streams that rise with us run into the Thames at Weybridge, and so to London. Of course therefore there must be lower ground all the way from

Selborne to South Lambeth; the distance between which, all the windings and indentings of the streams considered, cannot be less than an hundred miles.

I am, &c.

Letter 61

SINCE the weather of a district is undoubtedly part of its natural history, I shall make no further apology for the four following letters,* which will contain many particulars concerning some of the great frosts and a few respecting some very hot summers, that have distinguished themselves from the rest during the course of my observations.

As the frost in January 1768 was, for the small time it lasted, the most severe that we had then known for many years, and was remarkably injurious to evergreens, some account of its rigour, and reason of its ravages, may be useful, and not unacceptable to persons that delight in planting and ornamenting; and may particularly become a work that professes never to lose sight of utility.*

For the last two or three days of the former year there were considerable falls of snow, which lay deep and uniform on the ground without any drifting, wrapping up the more humble vegetation in perfect security. From the first day to the fifth of the new year more snow succeeded; but from that day the air became entirely clear; and the heat of the sun about noon had a considerable influence in sheltered situations.

It was in such an aspect that the snow on the author's evergreens was melted every day, and frozen intensely every night; so that the laurustines, bays, laurels, and arbutuses looked, in three or four days, as if they had been burnt in the fire; while a neighbour's plantation of the same kind, in a high cold situation, where the snow was never melted at all, remained uninjured.

From hence I would infer that it is the repeated melting and freezing of the snow that is so fatal to vegetation, rather than the severity of the cold. Therefore it highly behoves every planter, who wishes to escape the cruel mortification of losing in a few days the labour and hopes of years, to bestir himself on such emergencies; and, if his plantations are small, to avail himself of mats, cloths, peas-haum, straw, reeds, or any such covering, for a short time; or, if his shrubberies are extensive, to see that his people go about with prongs and forks, and carefully dislodge the snow from the boughs: since the naked foliage will shift much better for itself, than where the snow is partly melted and frozen again.

It may perhaps appear at first like a paradox; but doubtless the more tender trees and shrubs should never be planted in hot aspects; not only for the reason assigned above, but also because, thus circumstanced, they are disposed to shoot earlier in the spring, and to grow on later in the autumn, than they would otherwise do, and so are sufferers by lagging or early frosts. For this reason also plants from Siberia will hardly endure our climate; because, on the very first advances of spring, they shoot away, and so are cut off by the severe nights of March or April.

Dr Fothergill and others have experienced the same inconvenience with respect to the more tender shrubs from North-America; which they therefore plant under north-walls.* There should also perhaps be a wall to the east to defend them from the piercing blasts from that quarter.

This observation might without any impropriety be carried into animal life; for discerning bee-masters now find that their hives should not in the winter be exposed to the hot sun, because such unseasonable warmth awakens the inhabitants too early from their slumbers; and, by putting their juices into motion too soon, subjects them afterwards to inconveniencies when rigorous weather returns.

The coincidents attending this short but intense frost were, that the horses fell sick with an epidemic distemper, which injured the winds of many, and killed some; that colds and coughs were general among the human species; that it froze under people's beds for several nights;* that meat was so hard frozen that it could not be spitted, and could not be secured but in cellars; that several redwings and thrushes were killed by the frost; and that the large titmouse continued to pull straws lengthwise from the eaves of thatched houses and barns in a most adroit manner, for a purpose that has been explained already.[30]

On the 3d of January Benjamin Martin's thermometer within doors, in a close parlour where there was no fire, fell in the night to 20, and on the 4th to 18, and on the 7th to 17½, a degree of cold which the owner never since saw in the same situation; and he regrets much that he was not able at that juncture to attend his instrument abroad. All this time the wind continued north and north-east; and yet on the 8th roost-cocks, which had been silent, began to sound their clarions, and crows to clamour, as prognostic of milder weather; and, moreover, moles began to heave and work; and a manifest thaw took place.* From the latter circumstance we may conclude that thaws often originate under ground from warm vapours which arise; else how should subterraneous animals receive such early intimations of their approach. Moreover, we have often observed that cold seems to descend from above; for, when a thermometer hangs abroad in a frosty night, the intervention of a cloud shall immediately raise the mercury ten degrees; and a clear sky shall again compel it to descend to its former gage.

And here it may be proper to observe, on what has been said above, that though frosts advance to their utmost severity by somewhat of a regular gradation, yet thaws do not usually come on by as regular a declension of cold; but often take place immediately from intense freezing, as men in sickness often mend at once from a paroxysm.

To the great credit of Portugal laurels and American junipers, be it remembered that they remained untouched amidst the general havock: hence men should learn to orna-ment chiefly with such trees as are able to withstand accidental severities, and not subject themselves to the vexation of a loss which may befall them once perhaps in ten years, yet may hardly be recovered through the whole course of their lives.

As it appeared afterwards, the ilexes were much injured, the cypresses were half destroyed, the arbutuses lingered on, but never recovered, and the bays, laurustines, and laurels, were killed to the ground; and the very wild hollies, in hot aspects, were so much affected that they cast all their leaves.

By the 14th of January the snow was entirely gone; the turnips emerged not damaged at all, save in sunny places; the wheat looked delicately, and the garden plants were well preserved; for snow is the most kindly mantle that infant vegetation can be wrapped in: were it not for that friendly meteor, no vegetable life could exist at all in northerly regions. Yet in Sweden the earth in April is not divested of snow for more than a fortnight before the face of the country is covered with flowers.

Letter 62

THERE were some circumstances attending the remarkable frost of January 1776 so singular and striking, that a short detail of them may not be unacceptable.

The most certain way to be exact will be to copy the passages from my journal, which were taken from time to time as things occurred. But it may be proper previously to remark that the first week in January was uncommonly wet, and drowned with vast rains from every quarter: from whence may be inferred, as there is great reason to believe is the case, that intense frosts seldom take place till the earth is completely glutted and chilled with water;[31] and hence dry autumns are seldom followed by rigorous winters.

January 7th—Snow driving all the day, which was followed by frost, sleet, and some snow, till the 12th, when a prodigious mass overwhelmed all the works of men, drifting over the tops of the gates and filling the hollow lanes.

On the 14th the writer was obliged to be much abroad; and thinks he never before or since has encountered such rugged Siberian weather. Many of the narrow roads were now filled above the tops of the hedges; through which the snow was driven into most romantic and grotesque shapes, so striking to the imagination as not to be seen without wonder and pleasure. The poultry dared not to stir out of their roosting-places; for cocks and hens are so dazzled and confounded by the glare of snow that they would soon perish without assistance. The hares also lay sullenly in their seats, and would not move till compelled by hunger; being conscious, poor animals, that the drifts and heaps treacherously betray their footsteps, and prove fatal to numbers of them.

From the 14th the snow continued to increase, and began to stop the road waggons and coaches, which could no longer keep on their regular stages; and especially on the western roads, where the fall appears to have been deeper than in the south. The company at Bath, that wanted to attend the Queen's birth-day, were strangely incommoded: many carriages of persons, who got in their way to town from Bath as far as Marlborough, after strange embarrassments, here met with a *ne plus ultra* [an impasse]. The ladies fretted, and offered large rewards to labourers if they would shovel them a track to London: but the relentless heaps of snow were too bulky to be removed; and so the 18th passed over, leaving the company in very uncomfortable circumstances at the Castle and other inns.

On the 20th the sun shone out for the first time since the frost began; a circumstance that has been remarked before much in favour of vegetation. All this time the cold was not very intense, for the thermometer stood at 29, 28, 25, and thereabout; but on the 21st it descended to 20. The birds now began to be in a very pitiable and starving condition. Tamed by the season, sky-larks settled in the streets of towns, because they saw the ground was bare; rooks frequented dunghills close to houses; and crows watched horses as they passed, and greedily devoured what dropped from them; hares now came into men's gardens, and, scraping away the snow, devoured such plants as they could find.

On the 22nd the author had occasion to go to London through a sort of Laplandian-scene, very wild and grotesque indeed. But the metropolis itself exhibited a still more singular appearance than the country; for, being bedded deep in snow, the pavement of the streets could not be touched by the wheels or the horses' feet, so that the carriages ran about without the least noise. Such an exemption from din and clatter was strange, but not pleasant; it seemed to convey an uncomfortable idea of desolation:

> . . . ipsa silentia terrent.
> [the very silence terrifies.
> Virgil, *Aeneid* 2]

On the 27th much snow fell all day, and in the evening the frost became very intense. At South Lambeth, for the four following nights, the thermometer fell to 11, 7, 6, 6; and at Selborne to 7, 6, 10; and on the 31st of January, just before sun-rise, with rime on the trees and on the tube of the glass, the quicksilver sunk exactly to zero, being 32 degrees below the freezing point: but by eleven in the morning, though in the shade, it sprung up to 16½—a most unusual degree of cold this for the south of England![32] During these four nights the cold was so penetrating that it occasioned ice in warm chambers and under beds; and in the day the wind was so keen that persons of robust constitutions could scarcely endure to face it. The Thames was at once so frozen over both above and below bridge that crowds ran about on the ice. The streets were now strangely encumbered with snow, which crumbled and trod dusty; and, turning grey, resembled bay-salt: what had fallen on the roofs was so perfectly dry that, from first to last, it lay twenty-six days on the houses in the city; a longer time than had been remembered by the oldest housekeepers living. According to all appearances we might now have expected the continuance of this rigorous weather for weeks to come, since every night increased in severity; but behold, without any apparent cause, on the 1st of February a thaw took place, and some rain followed before night; making good the observation above, that frosts often go off as it were at once, without any gradual declension of cold. On the 2d of February the thaw persisted; and on the 3d swarms of little insects were frisking and sporting in a court-yard at South Lambeth, as if they had felt no frost. Why the juices in the small bodies and smaller limbs of such minute beings are not frozen is a matter of curious inquiry.

Severe frosts seem to be partial, or to run in currents; for, at the same juncture, as the author was informed by accurate correspondents, at Lyndon, in the county of Rutland, the thermometer stood at 19; at Blackburn, in Lancashire, at 19; and at Manchester at 21, 20, and 18. Thus does some unknown circumstance strangely overbalance latitude, and render the cold sometimes much greater in the southern than the northern parts of this kingdom.

The consequences of this severity were, that in Hampshire, at the melting of the snow, the wheat looked well, and the turnips came forth little injured. The laurels and laurustines were somewhat damaged, but only in hot aspects. No evergreens were quite destroyed, and not half the damage sustained that befell in January 1768. Those laurels that were a little scorched on the south-sides were perfectly untouched on their north-sides. The care taken to shake the snow day by day from the branches seemed greatly to avail the author's evergreens. A neighbour's laurel-hedge, in a high situation, and facing to the north, was perfectly green and vigorous;* and the Portugal laurels remained unhurt.

As to the birds, the thrushes and blackbirds were mostly destroyed; and the partridges, by the weather and poachers, were so thinned that few remained to breed the following year.

Letter 63

As the frost in December 1784 was very extraordinary, you, I trust, will not be displeased to hear the particulars; and especially when I promise to say no more about the severities of winter after I have finished this letter.

The first week in December was very wet, with the barometer very low. On the 7th, with the barometer at 28—five tenths, came on a vast snow, which continued all that day and the next, and most part of the following night; so that by the morning of the 9th the works of men were quite overwhelmed, the lanes filled so as to be impassable, and the ground covered twelve or fifteen inches without any drifting. In the evening of the 9th the air began to be so very sharp that we thought it would be curious to attend to the motions of a thermometer:

we therefore hung out two; one made by Martin and one by Dollond, which soon began to shew us what we were to expect; for, by ten o'clock, they fell to 21, and at eleven to 4, when we went to bed. On the 10th, in the morning, the quicksilver of Dollond's glass was down to half a degree below zero; and that of Martin's, which was absurdly graduated only to four degrees above zero, sunk quite into the brass guard of the ball; so that when the weather became most interesting this was useless.* On the 10th, at eleven at night, though the air was perfectly still, Dollond's glass went down to one degree below zero! This strange severity of the weather made me very desirous to know what degree of cold there might be in such an exalted and near situation as Newton. We had therefore, on the morning of the 10th, written to Mr——, and entreated him to hang out his thermometer, made by Adams; and to pay some attention to it morning and evening; expecting wonderful phaenomena, in so elevated a region, at two hundred feet or more above my house. But, behold! on the 10th, at eleven at night, it was down only to 17, and the next morning at 22, when mine was at ten! We were so disturbed at this unexpected reverse of comparative local cold, that we sent one of my glasses up, thinking that of Mr—— must, some how, be wrongly constructed. But, when the instruments came to be confronted, they went exactly together: so that, for one night at least, the cold at Newton was 18 degrees less than at Selborne; and, through the whole frost, 10 or 12 degrees; and indeed, when we came to observe consequences, we could readily credit this; for all my laurustines, bays, ilexes, arbutuses, cypresses, and even my Portugal laurels,[33] and (which occasions more regret) my fine sloping laurel-hedge, were scorched up; while, at Newton, the same trees have not lost a leaf!

We had steady frost on to the 25th, when the thermometer in the morning was down to 10 with us, and at Newton only to 21. Strong frost continued till the 31st, when some tendency to thaw was observed; and, by January the 3d, 1785, the thaw was confirmed, and some rain fell.

A circumstance that I must not omit, because it was new to us, is that on Friday, December the 10th, being bright sunshine, the air was full of icy *spiculae*, floating in all directions, like atoms in a sun-beam let into a dark room. We thought them at first particles of the rime falling from my tall hedges; but were soon convinced to the contrary, by making our observations in open places where no rime could reach us. Were they watery particles of the air frozen as they floated; or were they evaporations from the snow frozen as they mounted?

We were much obliged to the thermometers for the early information they gave us; and hurried our apples, pears, onions, potatoes, &c. into the cellar, and warm closets; while those who had not, or neglected such warnings, lost all their store of roots and fruits, and had their very bread and cheese frozen.

I must not omit to tell you that, during those two Siberian days, my parlour-cat was so electric, that had a person stroked her, and been properly insulated, the shock might have been given to a whole circle of people.

I forgot to mention before, that, during the two severe days, two men, who were tracing hares in the snow, had their feet frozen; and two men, who were much better employed, had their fingers so affected by the frost, while they were thrashing in a barn, that a mortification followed, from which they did not recover for many weeks.

The frost killed all the furze and most of the ivy, and in many places stripped the hollies of all their leaves. It came at a very early time of the year, before old November ended; and yet may be allowed from its effects to have exceeded any since 1739–40.

Letter 64

As the effects of heat are seldom very remarkable in the northerly climate of England, where the summers are often so defective in warmth and sun-shine as not to ripen the fruits of the earth so well as might be wished, I shall be more concise in my account of the severity of a summer season, and so make a little amends for the prolix account of the degrees of cold, and the inconveniencies that we suffered from some late rigorous winters.

The summers of 1781 and 1783 were unusually hot and dry; to them therefore I shall turn back in my journals, without recurring to any more distant period. In the former of these years my peach and nectarine-trees suffered so much from the heat that the rind on the bodies was scalded and came off; since which the trees have been in a decaying state. This may prove a hint to assiduous gardeners to fence and shelter their wall-trees with mats or boards, as they may easily do, because such annoyance is seldom of long continuance. During that summer also, I observed that my apples were coddled, as it were, on the trees; so that they had no quickness of flavour, and would not keep in winter. This circumstance put me in mind of what I have heard travellers assert, that they never ate a good apple or apricot in the south of Europe, where the heats were so great as to render the juices vapid and insipid.

The great pests of a garden are wasps, which destroy all the finer fruits just as they are coming into perfection. In 1781 we had none; in 1783 there were myriads; which would have devoured all the produce of my garden, had not we set the boys to take the nests,* and caught thousands with hazel-twigs tipped with bird-lime: we have since employed the boys to take and destroy the large breeding wasps in the spring. Such expedients have a great effect on these marauders, and will keep them under. Though wasps do not abound but in hot summers, yet they do not prevail in every hot summer, as I have instanced in the two years above-mentioned.

In the sultry season of 1783 honey-dews were so frequent as to deface and destroy the beauties of my garden. My honeysuckles, which were one week the most sweet and lovely objects that the eye could behold, became the next the most loathsome; being enveloped in a viscous substance, and loaded with black aphides, or smother-flies. The occasion of this clammy appearance seems to be this, that in hot weather the effluvia of flowers in fields and meadows and gardens are drawn up in the day by a brisk evaporation, and then in the night fall down again with the dews, in which they are entangled; that the air is strongly scented, and therefore impregnated with the particles of flowers in summer weather, our senses will inform us; and that this clammy sweet substance is of the vegetable kind we may learn from bees, to whom it is very grateful: and we may be assured that it falls in the night, because it is always first seen in warm still mornings.

On chalky and sandy soils, and in the hot villages about London, the thermometer has been often observed to mount as high as 83 or 84; but with us, in this hilly and woody district, I have hardly ever seen it exceed 80; nor does it often arrive at that pitch. The reason, I conclude, is, that our dense, clayey soil, so much shaded by trees, is not so easily heated through as those above-mentioned: and, besides, our mountains cause currents of air and breezes; and the vast effluvia from our woodlands temper and moderate our heats.

Letter 65

THE summer of the year 1783 was an amazing and portentous one, and full of horrible phaenomena; for, besides the alarming meteors and tremendous thunder-storms that affrighted and distressed the different counties of this kingdom, the peculiar haze, or smokey fog, that prevailed for many weeks in this island, and in every part of Europe, and even beyond its limits, was a most extraordinary appearance, unlike any thing known within the memory of man.* By my journal I find that I had noticed this strange occurrence from June 23 to July 20 inclusive, during which period the wind varied to every quarter without making any alteration in the air. The sun, at noon, looked as black as a clouded moon, and shed a rust-coloured ferruginous light on the ground, and floors of rooms; but was particularly lurid and blood-coloured at rising and setting. All the time the heat was so intense that butchers' meat could hardly be eaten on the day after it was killed; and the flies swarmed so in the lanes and hedges that they rendered the horses half frantic, and riding irksome. The country people began to look with a superstitious awe at the red, louring aspect of the sun; and indeed there was reason for the most enlightened person to be apprehensive; for, all the while, Calabria and part of the isle of Sicily, were torn and convulsed with earthquakes, and about that juncture a volcano sprung out of the sea on the coast of Norway. On this occasion Milton's noble simile of the sun, in his first book of *Paradise Lost*, frequently occurred to my mind; and it is indeed particularly applicable, because, towards the end, it alludes to a superstitious kind of dread, with which the minds of men are always impressed by such strange and unusual phaenomena.

> As when the sun, new risen,
> Looks through the horizontal, misty air,
> Shorn of his beams; or, from behind the moon,
> In dim eclipse, disastrous twilight sheds
> On half the nations, and with fear of change
> Perplexes monarchs . . .

[Milton, Paradise Lost, I]

Letter 66

WE are very seldom annoyed with thunder-storms: and it is no less remarkable than true, that those which arise in the south have hardly been known to reach this village; for, before they get over us, they take a direction to the east or the west, or sometimes divide into two, and go in part to one of those quarters, and in part to the other; as was truly the case in the summer 1783, when, though the country round was continually harassed with tempests, and often from the south, yet we escaped them all; as appears by my journal of that summer. The only way that I can at all account for this fact—for such it is—is that, on that quarter, between us and the sea, there are continual mountains, hill behind hill, such as Nore-hill, the Barnet, Butser-hill, and Ports-down, which some how divert the storms, and give them a different direction. High promontories, and elevated grounds, have always been observed to attract clouds and disarm them of their mischievous contents, which are discharged into the trees and summits as soon as they come in contact with these turbulent meteors; while the humble vales escape, because they are so far beneath them.

But, when I say I do not remember a thunder-storm from the south, I do not mean that we never have suffered from thunder-storms at all; for on June 5th, 1784, the thermometer in the morning being at 64, and at noon at 70, the barometer at 29—six tenths one-half, and the wind north, I observed a blue mist, smelling strongly of sulphur, hanging along our sloping woods, and seeming to indicate that thunder was at hand. I was called in about two in the afternoon, and so missed seeing the gathering of the clouds in the north; which they who were abroad assured me had something uncommon in its appearance. At about a quarter after two the storm began in the parish of Hartley, moving slowly from north to south; and from thence it came over Norton-farm, and so to Grange-farm, both in this parish. It began with vast drops of rain, which were soon succeeded by round hail, and then by convex pieces of ice, which measured three inches in girth. Had it been as extensive as it was violent, and of any continuance (for it was very short), it must have ravaged all the neighbourhood. In the parish of Hartley it did some damage to one farm; but Norton, which lay in the center of the storm, was greatly injured; as was Grange, which lay next to it. It did but just reach to the middle of the village, where the hail broke my north windows, and all my garden-lights and hand-glasses, and many of my neighbours' windows.* The extent of the storm was about two miles in length and one in breadth. We were just sitting down to dinner; but were soon diverted from our repast by the clattering of tiles and the jingling of glass. There fell at the same time prodigious torrents of rain on the farms above-mentioned, which occasioned a flood as violent as it was sudden; doing great damage to the meadows and fallows, by deluging the one and washing away the soil of the other. The hollow lane towards Alton was so torn and disordered as not to be passable till mended, rocks being removed that weighed two hundredweight. Those that saw the effect which the great hail had on ponds and pools say that the dashing of the water made an extraordinary appearance, the froth and spray standing up in the air three feet above the surface. The rushing and roaring of the hail, as it approached, was truly tremendous.

Though the clouds at South Lambeth, near London, were at that juncture thin and light, and no storm was in sight, nor within hearing, yet the air was strongly electric; for the bells of an electric machine at that place rang repeatedly, and fierce sparks were discharged.*

When I first took the present work in hand, I proposed to have added an *Annus Historico-naturalis*, or The Natural History of the Twelve Months of the Year; which would have comprised many incidents and occurrences that have not fallen in my way to be mentioned in my series of letters;—but as Mr Aikin of Warrington has lately published somewhat of this sort,* and as the length of my correspondence has sufficiently put your patience to the test, I shall here take a respectful leave of you and natural history together;* And am,

With all due deference and regard,
Your most obliged,
And most humble servant,
GIL. WHITE

Selborne,
June 25, 1787.

Gilbert White's tomb

WHITE'S NOTES

LETTERS TO PENNANT

1 This spring produced, September 14, 1781, after a severe hot summer, and a preceding dry spring and winter, nine gallons of water in a minute, which is five hundred and forty in an hour, and twelve thousand nine hundred and sixty, or two hundred and sixteen hogsheads, in twenty-four hours, or one natural day. At this time many of the wells failed, and all the ponds in the vales were dry. [The measurement given here seems to have been made by W's brother, Thomas, who was at Selborne at this time; as a sometime merchant his interests were much more towards such calculations than were W's.]

2 This soil produces good wheat and clover. [Arthur Young, *Farmer's Kalendar* (1771) advised 'the good farmer . . . to sow no barley or oats without clover'.]

3 There may probably be also in the chalk itself that is burnt for lime a proportion of sand; for few chalks are so pure as to have none.

4 To surbed stone is to set it edgewise, contrary to the posture it had in the quarry, says Dr Plot, *Oxfordsh.*, p. 77. But surbedding does not succeed in our dry walls; neither do we use it so in ovens, though he says it is best for Teynton stone. [W is using the 2nd edn. of this book (first published in 1677). Plot advised surbedding to prevent the stone from 'flying' under pressure of frost or, in ovens, heat. Teynton, also Taynton, is near Burford.]

5 'Firestone is full of salts, and has no sulphur: must be close grained, and have no interstices. Nothing supports fire like salts; saltstone perishes exposed to wet and frost.' Plot's *Staff.* [1686], p. 152.

6 A very intelligent gentleman assures me (and he speaks from upwards of forty years' experience) that the mean rain of any place cannot be ascertained till a person has measured it for a very long period. 'If I had only measured the rain,' says he, 'for the four first years, from 1740 to 1743, I should have said the mean rain at Lyndon was 16½ inches for the year; if from 1740 to 1750, 18½ inches. The mean rain before 1763 was 20¼; from 1763 and since 25½; from 1770 to 1780, 26. If only 1773, 1774, and 1775, had been measured, Lyndon mean rain would have been called 32 inches.' [The data came from W's brother-in-law, Thomas Barker of Lyndon Hall, Rutland. He was the principal

agent of W's interest in meteorology, and they shared a youthful interest in natural history from as early as 1736.]

7   A STATE *of the* Parish *of* SELBORNE, *taken* October 4, 1783.
      The number of tenements or families, 136.

The number of inhabitants in⎫ Total 676; near five inhabi-
   the street is 313   ⎬ tants to each tenement.
In the rest of the parish 363 ⎭

In the time of the Rev. *Gilbert White*, Vicar, who died in 1727–8, the number of inhabitants was computed at about 500. [This Gilbert White was W's paternal grandfather, who held office at Selborne from 1681. The date of his death is not as ambiguous as it appears: the New Year began on 1 Jan. only after 1752, whereas formerly it began 25 Mar.]

Average of baptisms for 60 years.

From 1720 to 1729,⎫ Males 6.9   ⎫   From 1740 to⎫ M. 9.2 ⎫
   both years inclus.⎬ Females 6.0 ⎬ 12.9   1749, incl.⎬ F. 6.6 ⎬ 15.8
From 1730 to 1739,⎫ Males 8.2   ⎫   From 1750 to⎫ M. 7.6 ⎫
   both years inclus.⎬ Females 7.1 ⎬ 15.3   1759, incl.⎬ F. 8.1 ⎬ 15.7

                From 1760 to⎫ M. 9.1 ⎫
                   1769, incl.⎬ F. 8.9 ⎬ 18.0
                From 1770 to⎫ M. 10.5 ⎫
                   1779, incl.⎬ F. 9.8 ⎬ 20.3

        Total of baptisms of Males 515 ⎫ 980
                              Females 465 ⎭

     Total of baptisms from 1720 to 1779, both inclusive:
                60 years 980

Average of burials for 60 years.

From 1720 to 1729,⎫ Males 4.8   ⎫   From 1740 to⎫ M. 4.6 ⎫
   both years inclus.⎬ Females 5.1 ⎬ 9.9   1749, incl.⎬ F. 3.8 ⎬ 8.4
From 1730 to 1739,⎫ Males 4.8   ⎫   From 1750 to⎫ M. 4.9 ⎫
   both years inclus.⎬ Females 5.8 ⎬ 10.6   1759, incl.⎬ F. 5.1 ⎬ 10.0

                From 1760 to⎫ M. 6.9 ⎫
                   1769, incl.⎬ F. 6.5 ⎬ 13.4
                From 1770 to⎫ M. 5.5 ⎫
                   1779, incl.⎬ F. 6.2 ⎬ 11.7

        Total of burials of Males 315 ⎫ 640
                             Females 325 ⎭

     Total of burials from 1720 to 1779, both inclusive:
                60 years 640

Baptisms exceed burials by more than one third.
Baptisms of Males exceed Females by one tenth, or one in ten.
Burials of Females exceed Males by one in thirty.

It appears that a child, born and bred in this parish, has an
equal chance to live above forty years.
Twins thirteen times, many of whom dying young have lessened
the chance for life.
Chances for life in men and women appear to be equal.

A TABLE *of the* Baptisms, Burials, *and* Marriages, *from* January
2, 1761, *to* December 25, 1780, *in the* Parish *of* SELBORNE.

	BAPTISMS			BURIALS			MAR.
	M.	F.	Tot.	M.	F.	Tot.	
1761	8	10	18	2	4	6	3
1762	7	8	15	10	14	24	6
1763	8	10	18	3	4	7	5
1764	11	9	20	10	8	18	6
1765	12	6	18	9	7	16	6
1766	9	13	22	10	6	16	4
1767	14	5	19	6	5	11	2
1768	7	6	13	2	5	7	6
1769	9	14	23	6	5	11	2
1770	10	13	23	4	7	11	3
	95	94	189	62	65	127	43

	BAPTISMS			BURIALS			MAR.
	M.	F.	Tot.	M.	F.	Tot.	
1771	10	6	16	3	4	7	4
1772	11	10	21	6	10	16	3
1773	8	5	13	7	5	12	3
1774	6	13	19	2	8	10	1
1775	20	7	27	13	8	21	6
1776	11	10	21	4	6	10	6
1777	8	13	21	7	3	10	4
1778	7	13	20	3	4	7	5
1779	14	8	22	5	6	11	5
1780	8	9	17	11	4	15	3
	103	94	197	61	58	119	40
	95	94	189	62	65	127	43
	198	188	386	123	123	246	83

During this period of twenty years the birth of males exceeded those of females by 10

The burials of each sex were equal.

And the births exceeded the deaths by 140

[At first sight W's interest in the parish population is a matter of computation and, in consequence, speculation about life expectancy. Yet, underlying the tabulated data is the question of 'Divine Wisdom and Providence'. Derham, one of W's religious mentors, devoted a whole chapter of his *Physico-Theology* (1713) to 'Of the Balance of Animals, or their due Proportion wherewith the World is stocked' and raised fundamental questions about the numerical relationship of males to females and of births to deaths. That in England, generally, there were 13 females to 13.7 males was seen as a 'Work of the Divine Providence, and not a Matter of Chance' since the inequality showed (*a*) that one man should have only one wife; (*b*) that every woman might have a husband; and (*c*) that the 'Surplusage of Males [was] very useful for the Supplies of War, the Seas, and other such expences of the Men above the Women'. As for births and deaths, Derham evidenced that 'fewer Die than are Born, there being but 1 Death to $1^{12}/_{100}$ Births' and advanced this proportion as 'an admirable Provision for the extraordinary Emergencies and Occasions of the World; to supply unhealthful Places, where Death out-runs Life; to make up the Ravages of great Plagues, and Disease ... and to afford a sufficient Number for Colonies in the unpeopled Parts of the Earth'.

The data of the parish, W must have felt, confirmed the weighting of Derham's values: Selborne, too, observed the contours of Divine Providence.]

8 Since the passage above was written, I am happy in being able to say that the spinning employment is a little revived, to the no small comfort of the industrious housewife.

9 See his Hist. of *Staffordshire*. [In fact, Plot's claim was no more than that no one had 'ever heard of them ('subterraneous trees') in the vales of Evesham or Aylesbury ... the vales of white or red horse' (*Staffordshire* (1686), 216). Interest in the phenomenon lay not only in the occurrence but, since trees do not grow underground, in what possessed men to cut down fine timber and bury it; or, as Plot puts it, in what were the 'causes assignable of such wast of timber, and the sepulture of it' (p. 220).]

10 Old people have assured me, that on a winter's morning they have discovered these trees, in the bogs, by the hoar frost, which lay longer over the space where they were concealed, than on the surrounding morass. Nor does this seem to be a fanciful notion, but consistent with true philosophy. Dr Hales saith, 'That the warmth of the earth, at some depth under ground, has an influence in promoting a thaw, as well as the change of the weather from a freezing to a thawing state, is manifest, from this observation, viz. Nov. 29, 1731, a little snow having fallen in the night, it was, by eleven the next morning, mostly melted away on the surface of the earth, except in several places in Bushy park [between Teddington and Hampton Court], where there were drains dug and covered with earth, on which the snow continued to lie, whether those drains were full of water or dry; as also where elm-pipes lay under ground: a plain proof this, that those drains intercepted the warmth of the earth from ascending from greater depths below them: for the snow lay where the drain had more than four feet of earth over it. It continued also to lie on thatch, tiles, and the tops of walls.' See Hales's *Haemastatics*: p. 360. *Quere*: Might not such observations be reduced to domestic use, by promoting the discovery of old obliterated drains and wells about houses; and in Roman stations and camps lead to the finding of pavements, baths and graves, and other hidden relics of curious antiquity?

11 Statute 9 Geo. I. c. 22. [Criminal Law statute, passed in 1722.]

12 This chase remains un-stocked to this day; the bishop was Dr Hoadly. [Benjamin Hoadly (1676–1761) was translated from Salisbury to Winchester, his fourth bishopric, in 1734. The chase was the hunting ground of bishops of Winchester attached to the medieval episcopal palace at Bishop's Waltham, Hampshire.]

13 For this privilege the owner of that estate used to pay to the king annually seven bushels of oats.

14 In The Holt, where a full stock of fallow-deer has been kept up till lately, no sheep are admitted to this day.

15 I mean that sort which, rising into tall hassocks, is called by the foresters torrets; a corruption, I suppose of turrets [greater tussock-sedge, *Carex paniculata*].

16 For which consult letter 42 to Mr Barrington.

17 In the beginning of the summer 1787 the royal forests of Wolmer and Holt were measured by persons sent down by government. [Details of this survey are in a *Report* of the Commissioners of

Woods and Forests; Wolmer was valued at 4 shillings (20 pence) with the Holt at up to 30 shillings (£1.50).]

18 'In *Rot. Inquisit. de statu* forest, in *Scaccar.* 36 Ed. 3. it is called Aisholt [In the Rolls of Inquiries concerning the state of forest-land in the Exchequer, 36th year of K. Edward III (1362–3), it is called Aisholt].' In the same, '*Tit.* Woolmer and Aisholt *Hantisc. Dominus Rex habet unam capellam in haia sua de Kingesle* [Title: Woolmer and Aisholt, Hampshire. The Lord King has one chapel in his park at Kingsley].' '*Haia, sepes, sepimentum, parcus: a Gall. haie* and *haye.*'—Spelman's *Glossary.* [Henry Spelman (d. 1641) prepared a *Glossary* of law terms (vol. i, 1626; vol. ii, 1664, posth.). W cites a number of terms for kinds of park fencing—*haia (a Gall:* from the French) indicating a living hedge, with *sepes* (palings) and *sepimentum* (woven or plaited fencing).]

19 This prince was the inventor of mezzotinto. [In fact Prince Rupert only introduced the art to England, the inventor being the Dutch-German Ludwig von Siegen (1609–*c.* 1680).]

20 This hawk proved to be the *falco peregrinus*; a variety. [W has been reproached for not identifying the peregrine but although it was more widely distributed in his time than in ours (and in southern England), its customary habitat is cliffs and moorland, a territory with which W was unfamiliar.]

21 See Adanson's *Voyage to Senegal* [published in English in 1759; W is responding to Adanson's report that when he stayed (Feb. 1750) in a hut on the Gambia 'a great number of our European swallows resorted hither every evening, and passed the night upon the rafters; for . . . they do not build nests in this country, but only come to spend the winter'].

22 See Ray's *Travels* [1738 (1673)], p. 466.

23 In answer to this account, Mr Pennant sent me the following curious and pertinent reply. 'I was much surprised to find in the antelope something analogous to what you mention as so remarkable in deer. This animal also has a long slit beneath each eye, which can be opened and shut at pleasure. On holding an orange to one, the creature made as much use of those orifices as of his nostrils, applying them to the fruit, and seeming to smell it through them.'

24 *Brit. Zool.* edit. 1776, octavo, p. 381.

25 James 3: 7. [Authorized Version of the Bible (1611)].

26 Cressi-hall is near Spalding, in Lincolnshire. [The seat of the Heron family, the hall was destroyed by fire in 1791 and the heronry dispersed. Pennant visited in June 1769 at the start of

his tour to Scotland; the error to which W refers is Pennant's discovery that the crested and uncrested herons, which he described in *BZ* (vol. ii; 1768) as two species were examples of sexual variation in a single species.]

27 For this *salicaria* see letter August 30, 1769 [TP 25].

28 The angler's may-fly, the *ephemera vulgata* Linn. comes forth from its aurelia state, and emerges out of the water about six in the evening, and dies about eleven at night, determining the date of its fly state in about five or six hours. They usually begin to appear about the 4th of June, and continue in succession for near a fortnight. See Swammerdam, Derham, Scopoli, &c.

29 Vagrant cuckoo; so called because, being tied down by no incubation or attendance about the nutrition of its young, it wanders without control.

30 *Charadrius oedicnemus* [*Burhinus oedicnemus*].

31 *Gryllus campestris*.

32 In hot summer nights woodlarks soar to a prodigious height and hang singing in the air.

33 The light of the female glow-worm (as she often crawls up the stalk of a grass to make herself more conspicuous) is a signal to the male, which is a slender dusky *scarabaeus* [*Lampyris noctiluca*].

34 See the story of Hero and Leander [a classical allusion praised by a contemporary of W's (Joseph Warton, head of Winchester) as 'the happiest . . . in our language'. But W's poetry runs away with him: both sexes of glow-worm are luminous, although the female more brightly so than the male; but then the male is winged and can fly whereas the female is solely ambulant].

35 The little bat appears almost every month in the year; but I have never seen the large ones till the end of April, nor after July. They are most common in June, but never in any plenty: are a rare species with us [so rare that W's description (of the noctule or great bat, *Nyctalus noctule*) was the first in English; the 'little bat' is the common pipistrelle].

36 *Annus Primus Historico-Naturalis* [1769].

37 See his *Elenchus vegetabilium et animalium per Austriam inferiorem*, &c. [1756].

38 See letter 53 to Mr Barrington.

39 [For this bird and for each species mentioned in twenty of the following twenty-two paras. (no reference being given for the 'house-swallow' or for 'Wagtails'), W provides a reference to the relevant page in Pennant's *British Zoology*. The pages cited,

beginning with that for the osprey, are: 128, 161, 167, 198, 216, 224, 229 (all in vol. i., 1768); 237, 242, 244, 245, 271–2 (for W's '270, 271'), 269, 300, 306, 358, 360, 409, 475 (in vol. ii, 1768); 15, 16 (in vol. iv., 1770).]

40   They eat also the berries of the ivy, the honey-suckle, and the *euonymus europaeus*, or spindle-tree.

41   Sweden 221; Great-Britain 252 species. [Numbers of this kind are dependent absolutely on definition: today, if rarities are excluded the number of species in the British Isles has scarcely changed.]

42   See Derham's *Physico-Theology* [1713], p. 235. [W draws attention here not to a list of species but to Derham's account of the artifice of varied kinds of caddis larvae (species of *Phryganea*) that construct protective cases with 'a notable architectonick Faculty . . . some to be heavier than Water, that the Animal may remain at the bottom, where its Food is, for which purpose they use Stones . . . and some to be lighter than Water, to float on the Top'.]

43   Some old sportsmen say that the main part of these flocks used to withdraw as soon as the heavy Christmas frosts were over.

LETTERS TO BARRINGTON

1   Job 39: 16, 17. [W omits from this biblical quotation the second part of v. 16: 'her labour is in vain without fear'. The verses are part of a poetic account, attributed by the writer of the book of Job to God, of the mysteries of creation and are designed to demonstrate man's impotence and ignorance in the face of God's omnipotence and omniscience.]

2   This work he calls his *Annus Primus Historico-Naturalis* [1769].

3   See letter 25 to Mr Pennant.

4   See letter 42 to Mr Barrington.

5   I have read a like anecdote of a swan.

6   Isaiah 1: 3. [W is often charged with the temper of a provincial, unaware of national and foreign concerns; but here, in quoting as he does, he might equally be praised for a prophetic prescience. The verse cited continues: 'but Israel doth not know, my people doth not consider [who their master is]', for within a matter of months the Boston Tea Party (1773) was to herald the American War of Independence.]

7   See Ulloa's *Travels*. [Antonio de Ulloa, *A Voyage to South-America*,

transl. J. Adams (1758), was a popular text and reprinted in 1760 and 1772. Guayaquil is a coastal city in Ecuador.]

8   Mr Courthope of Danny [near Hassocks, East Sussex. Ray dedicated to Courthope his *Collection of English Words Not Generally Known* (1673)].

9   Sir Ashton Lever's Museum. [The museum, opened in 1774 at Leicester House, London, was the delight of its founder, Ashton Lever (1729–88) of Accrington Hall, Lancashire. The collection, embracing every kind of curiosity and valued at £53,000 by a parliamentary committee in 1783, started life in the 1750s as an aviary of live birds.]

10   Nigra velut magnas domini cum divitis aedes
     Pervolat, et pennis alta atria lustrat hirundo,
     Pabula parva legens, nidisque loquacibus escas:
     Et nunc porticibus vacuis, nunc humida circum
     Stagna sonat.

[As when a blue-black swallow sweeps through the great mansion of a rich lord and flies amidst its lofty rooms, seizing scraps of sustenance for her noisy nestlings, so can we hear her twittering calls echo now through the spacious entranceways, now over the canals of the water-gardens. *Aeneid*, 12.]

11   John Antony Scopoli, of Carniola, M.D. [see Biographical Notes].

12   Tobit 2: 10. [A book of the Apocrypha, the verse cited reading in part: 'there were sparrows in the wall, and mine eyes being open, the sparrows muted warm dung into mine eyes'.]

13   See Bell's *Travels in China*. [John Bell (1691–1780) travelled extensively in Asia as part of a mission from Peter the Great, his journey to Peking lasting from 1718 to 1722. The *Travels* were published in 1763.]

14   A besom of this sort is to be seen in Sir Ashton Lever's Museum. [The besom was a gift from W to Barrington, who then presented it to the museum—see in *Selborne* MS unpublished portion of letter to Barrington of 13 May 1778.]

15   For a similar practice, see Plot's *Staffordshire* [1686]. [Plot's account gives some graphic detail: he tells of workmen felling a solid oak who 'perceived blood to follow the saw'. After completing the sawing and splitting, they discovered they had cut through 'the body of a Hardi-shrew', although two others escaped alive (p. 222).]

16   *Vide* Kalm's *Travels to North-America* [transl. J. R. Forster (2nd

edn., 1772); Peter Kalm, a disciple of Linnaeus, journeyed to America in 1748 and returned to Europe in 1751].

17 *Histoire de l'Académie Royale*, 1752.

18 Farmer Young, of Norton farm [on the edge of the parish just north of Selborne], says that this spring (1777) about four acres of his wheat in one field was entirely destroyed by slugs, which swarmed on the blades of corn, and devoured it as fast as it sprang.

19 See Leviticus 13 and 14 [these chapters detail the recognition of the disease and the procedures for cleansing and thanksgiving].

20 Viz. Six hundred bacons, eighty carcasses of beef, and six hundred muttons.

21 'In monasteries the lamp of knowledge continued to burn, however dimly. In them men of business were formed for the state: the art of writing was cultivated by the monks; they were the only proficients in mechanics, gardening, and architecture.'— See Dalrymple's *Annals of Scotland* [1776, 1779].

22 See the late Voyages to the south-seas [W refers to Cook's first voyage of 1768–71, accounts of which occurred in Parkinson (1773), Hawkesworth (1773), and to similar publications about his second journey of 1772–5—e.g. Marra (1775), Forster (1778)].

23 See *Spectator*, Vol. VII. No. 512 [the *Spectator* essay of 17 Oct. 1712 advocates counsel by fable: the wisdom of the owls, as reported by a vizier to his sultan, so moved the heart of this prince of war that he not only abandoned a career of pillage and devastation but rebuilt what he had previously destroyed].

24 We have observed that they cast these skins in April, which are then seen lying at the mouths of their holes.

25 Exod. 8: 3.

26 For various methods by which several insects shift their quarters, see Derham's *Physico-Theology*. [W's imagination was caught not only by Derham's account of gossamer, but also by his speculation that some insects had the 'Faculty of inflating their Bodies . . . and making [them] buoyant, and lighter than Air . . . or the Eggs of such as are oviparous, may be light enough to float in the Air'.]

27 See my tenth and eleventh letter to that gentleman.

28 Hasselquist [a disciple of Linnaeus], in his *Travels to the Levant* [1766], observes that the dogs and vultures at Grand Cairo maintain such a friendly intercourse as to bring up their young together in the same place.

29 The Chinese word for a dog to an European ear sounds like *quihloh*. [Someone may have been having fun with W's informant, Charles Etty, since *quihloh* means 'foreign devil'.]

30 See Letter 41 to Mr Pennant.

31 The autumn preceding January 1768 was very wet, and particularly the month of September, during which there fell at Lyndon, in the county of Rutland, six inches and an half of rain. And the terrible long frost in 1739–40 set in after a rainy season, and when the springs were very high.

32 At Selborne the cold was greater than at any other place that the author could hear of with certainty: though some reported at the time that at a village in Kent the thermometer fell two degrees below zero, viz. 34 degrees below the freezing point.

The thermometer used at Selborne was graduated by Benjamin Martin. [Martin (1704–82) was a London instrument-maker; as a young man he practised in Chichester.]

33 Mr Miller, in his *Gardener's Dictionary*, says positively that the Portugal laurels remained untouched in the remarkable frost of 1739–40. So that either that accurate observer was much mistaken, or else the frost of December 1784 was much more severe and destructive than that in the year above-mentioned.

EXPLANATORY NOTES

ABBREVIATIONS

Antiq.   Gilbert White, 'Antiquities', part II of the first edition of *Selborne* in 1789.

BZ   Thomas Pennant, *British Zoology* (1768–70).

GW   Paul G. M. Foster, *Gilbert White: A Scientific Biography* (London: Christopher Helm, 1988).

Johnson   Samuel Johnson, *A Dictionary of the English Language* (2 vols.; 1755).

NJ   'Naturalist's Journal' completed by White and now deposited at the British Library. Dates such as 19/20 May indicate observations recorded on an extra sheet added between those for 19 May and 20 May.

Phil. Trans.   *Philosophical Transactions.*

5   *epigraphs*: the first two appeared on the half-title page of the 1789 edition of *Selborne*. The quotation from the *Odyssey* may be translated: 'It is a rough country, but a good place for rearing young men. I for my part cannot envisage anything sweeter than one's own country.' The Cicero quotation runs: 'In short, the whole of that territory of ours is rough and mountainous, [but] dependable, trustworthy, and a protector of her own people.'

     epigraphs: the second two appeared on the title page of the 1789 edition of *The Natural History and Antiquities of Selborne*. The quotation from Horace runs: 'I, like a bee on Mount Matinus, industriously engage in a favourite activity . . .'. The Scopoli quotation runs: 'To attempt a description of everything there is in the world, whether of God's work or of the forces of natural creation, is a fruitless endeavour far beyond the skills, or even the lifetime, of one man. For this reason the most useful task is the preparation of books on the fauna and flora of a particular region: it is monographs that are of the greatest value.'

7   *stationary*: men who, by virtue of their circumstances, live for a long period in a single neighbourhood.

     Southampton: name for the administrative area until reorganization in the 1950s. W's use later of Hampshire refers to the geographical territory.

member . . . society: Richard Chandler (1738–1810), to whom W considered addressing his letters on the antiquities of the parish; his assistant was Ralph Churton. The President whom W acknowledges was George Horne.

11 *hanging*: the pendent branches of the beech trees on the slope of the hill hang down and appear to clothe it.

12 *cart-way*: the 'single straggling street' of the previous paragraph.

incongruous: contrasting. The 'rank clay' W goes on to describe is chalk marl; the black malm is the chloritic marl of the Upper Greensand. Later in the letter W distinguishes several other beds of the Upper Greensand series, concluding with the 'white soil' on which the 'brightest' (best) hops grow.

town: in earlier centuries Selborne held a regular market, but by W's time the sole activity of this kind was an annual fair on 29 May (see Antiq. 26).

13 *German ocean*: derived from the Latin author Tacitus, who named what is now called the North Sea *Oceanus Germanicus*; hence a term deployed in W's time by writers with a classical education.

limpid: clear—from the purifying action of the water filtering through the chalk.

the soil . . . roads: this is the area of gault, a grey loamy clay, between Temple and Blackdown which further north runs into the forest of Alice Holt (see TP 9).

will produce . . . turnips: lime to neutralize the acidity of the sands and to act as a binding agent, and turnips to provide nutrients and vegetative matter.

14 *a broad-leaved elm . . . 1703*: the Great Storm, from 26 Nov. to 1 Dec., is described as the worst England has endured—8,000 lives were lost in flooding. An absorbing period source is Defoe, *The Storm* (1704). The tree W describes was a Wych elm, *Ulmus glabra*.

Plestor: in Antiq. 10 W terms this public space '*locus ludorum*', a play area, and traces its name to the Saxon '*Plegestow*'; it had, and still has, the characteristics of a village green.

the vicar . . . died: the attempt to re-set the oak was by W's grandfather, also a Gilbert White, vicar at Selborne 1681–1727. The oak on the Plestor today was planted in 1897, Queen Victoria's Diamond Jubilee; and on the site of W's oak now stands a sycamore.

the bridge . . . near Hampton Court: the first bridge at this point, in chinoiserie style, was dated 1753; the Toy was an inn.

16   *zoophyte*: coined in the seventeenth century to describe natural objects that seemed partly animal (Gk. *zoon*, an animal) and partly vegetable (Gk. *phyton*, a plant).

my specimen . . . engraved: published by B. White & Son (W's bookseller brother and nephew) on 1 Nov. 1788 and included in *Selborne* (1789). The status of fossils in W's time was undetermined and this, together with the difficulties W met with in attempting to identify his fossil, explain the varied list of authorities: Martin Lister (?1638–1712), an English zoologist and correspondent of Ray; George Rumphius (1628–1702), a German; and Antoine-Joseph Dézallier d'Argenville (1680–1765), whose *La Conchyliologie* (1742) appeared in several later editions (he was also credited with being the first to describe, in *La Théorie et la pratique du jardinage* (1709), a ha-ha, a construction W built at Selborne (1759–61). W's investigations at Leicester-house, which displayed the Lever Museum (comprising 7,879 lots at its disposal over sixty-five days in 1806) only puzzled him more.

path: the Bostal, cut in 1780, which gave a gentler ascent to Selborne Common than the Zigzag of 1753.

recent production: evidently casts of ammonites, not fossilized shells. Clay's Pond is on the very edge of the parish, south-east of Well-head, and was a source of chalk marl spread on fields to be reduced by the action of winter weather and thereby enrich the soil.

17   *forest-stone*: Folkestone beds of the Lower Greensand, used, as W describes in his next paragraph, for galleting, a fairly common decoration on flint houses.

19   *filices*: ferns. As well as being struck by the form of ferns, W is intrigued that the damp shade of the hollow lanes encourages some plants to flourish.

all its kindly aspects: because of the hilly nature of the region, very varied natural habitats occur.

bounds: in a fenceless age and in a time when so much civil and ecclesiastical duty related to the parish, knowledge of the parish perimeter was regularly established by perambulation. The Selborne Registers record the bounds being walked in 1703, 1741, 1748, 1765, 1771, 1780: each occasion took three days.

effluvia: honey-dew, the excreta of aphids, although W took it to be the secretion of arboreal flowers mixed with dew. Cf. NJ June 1783.

37!: Exclamatory surprise that in eight months there was as much rain as might fall in a full year. W acquired a rain-measure only in April 1779.

20 *timber*: especially oaks, for the tan to cure hides into leather.

21 *aquatic tree*: W may be mistaken, as a forest Keeper claimed in 1789 that fir had been dug up.

teals: the *Selborne* MS provides here a W autograph note, 'See Letter 15 to Mr Barrington'.

unreasonable: as a young man W was well versed in field diversions himself and may be offering a private self-reproach for his own activities; but he also knew they possessed the potential for instruction—see DB 23.

shooting flying: the principal game, historically, had been deer and the ground-hugging partridge, but partly because of the decline in deer (through poaching) and partly because of improvements to the flintlock gun, the pheasant became, in the 1780s, the main target.

beautiful link . . . wanting: there is a suggestion here that W was conscious of man's role in maintaining the diversity of fauna in the parish: Providence, it seemed, could be abused.

22 *Waltham blacks*: poachers in Waltham Chase who blackened their faces and sometimes wore black gloves; their exploits would still have been a matter of report at the time W took up a curacy at Durley (near Bishop's Waltham, south Hampshire) some thirty years later.

riding-school: Philip Astley (1742–1814) was a flamboyant, Herculean figure. From 1769 he ran near Westminster Bridge an equestrian troupe under personal licence from George III, and toured, with great success, throughout the British Isles and on the Continent.

devoted: singled out as doomed to destruction, chosen as prey.

23 *ashes for their grasses*: W fertilized his own fields in this way— 'Sowed [scattered] 48 bushels of peat-ashes on the great meadow' (NJ 6 Feb. 1784). As for lime-burning, a later wet autumn (Oct., 1792) prompted: 'The brick-burner can get no dry heath to burn his lime, & bricks'.

24 *brouze*: grazing; and not for cattle alone, since the practice of burning became an important element in the management of heathland, and a regular source of conflict between landowners (who burned to provide new growth for pheasant) and freemen (whose grazing was thereby interrupted).

gentleman: Henry White, W's brother, clergyman at Fyfield, near Andover.

feast of St Barnabas: 11 June, the longest day before the calendar was changed in 1752 (when what would have been 3 Sept. became 14 Sept.).

26 *A various ... surface*: James Thomson, *The Seasons* (1746), 'Summer', ll. 485–9. At l. 486 W's omission is 'Rural Confusion!', a period phrase for the varied scenery he most enjoyed.

for ... sake: the three lakes (Hogmer, Cranmer, Wolmer) all contain the suffix 'mer' (Lat. *mare*, sea), meaning an area of standing water. The addition of 'pond' is therefore redundant but, at the same time, aggrandizes Wolmer, the largest of the three.

denominations: species, in this instance of a dozen or more.

27 *picturesque*: resembling a picture; a term that entered W's NJ about ten years earlier, esp. to describe views of fog seen from above—e.g. 'Beautiful picturesque, partial [patchy] fogs along the vales, representing rivers, islands, & arms of the sea!' (NJ 3 Dec. 1789).

district: the account of coins is given in Antiq. 1. They were found in the dry summers of 1740 and 1741, and were Roman of the second century AD; more were discovered in 1774, and in 1873 a cache of 30,000 coins (of the third century AD) was dug up.

Elmer: Stephen Elmer (d. 1796), maltster of Farnham, was elected an associate of the Royal Academy in 1772 and exhibited regularly for over twenty years. He was commissioned to paint a mule (a hybrid between a cock pheasant and a peahen) given to W by Lord Stawel in 1790; an engraving of this painting was first published, 14 Apr. 1795, as an illustration to W's *A Naturalist's Calendar* (1795, ed. John Aikin), and it was also included in the second (English) edn. of *Selborne* (1802).

28 *1784*: this date is a useful reminder that it is in the 1780s that W was actively preparing the first nine letters of *Selborne*; they serve as an introduction to his real correspondence, which, in an edited form, begins in 1767 with TP 10.

run: bleed with sap.

29 *1767*: this letter is an edited version of W's first letter to Pennant. The real letter, which is extant, is dated 10 August 1767.

childhood: W's first written observations on natural history are claimed to be notebook-entries made in 1736 when he was visiting relatives in Rutland.

person ... in Sussex: W's informant here is probably Henry Snooke, later clergyman at Ringmer, Sussex. The clergyman of the previous paragraph is most likely Charles White, an uncle.

30 *fieldfares*: not fieldfares, *Turdus pilaris*, but redwings, *T. musicus*, the smaller of our winter thrushes, although the two species often associate.

yellow bird ... woods: the bird with the 'sibilous shivering' song, often characterized as a grasshopper-like trill, is not the grasshopper warbler, *Locustella naevia*, but the wood warbler, *Phylloscopus sibilatrix*.

it takes ... together: this is a good example of W's interest in bird behaviour, both in its own right and, when described in detail, as a characteristic of a particular species and hence as a means of identification.

town: W travelled to London 18 Apr. and returned to Selborne on 12 June 1767. The meeting with Pennant occurred between these dates.

31 *non-descript*: a technical term relevant to the first stages of description of the world's flora and fauna; it means neither ordinary nor uninteresting but, literally, not yet described.

Spring-Gardens: the site also of the more well-known Vauxhall Gardens (Lambeth); W was ordained to the priesthood in the chapel there by the Bishop of Hereford on 11 March 1749.

abrupt ... country: terms of affection and approval that indicate W's delight in the ever-changing scenery and habitat of his native parish—and contrasting with his experience of fenland (Isle of Ely) where he stayed for six months in 1746.

32 *outlet*: an enclosure attached to a house; W's garden and orchard.

33 *squab-young*: newly-hatched birds, not yet feathered.

34 *frequent*: 'resort often to' (Johnson).

preserved in brandy: there were many recipes in W's time for an effective (and cheap) mode of preservation. Formaldehyde was not to be available for another 100 years.

36 *congeners*: birds of the same class, in this instance other *Fringillidae*, seed-eating birds.

Swedish naturalist: Alexander Berger, whose 'Flora' was first published in Stillingfleet (1762). W's observations of swallows on the aits occurred during visits to his lifelong friend, John Mulso, vicar at Sunbury-on-Thames 1746–59.

Borough: 'A town with a [mayor and] corporation' (Johnson); in this case Southwark. W's informant was probably his brother Thomas, a wholesale merchant.

the . . . birds: W means here species such as warblers, pipits, and flycatchers.

37 *such intelligence . . . continent*: to collect data and establish a co-ordinated understanding of migration and, hence, regional variation of fauna (esp. birds) was an interest W held throughout his life.

38 *those parts*: W travelled regularly into East Sussex to visit his aunt, Rebecca Snooke, whose husband was clergyman at Ringmer.

friend . . . Channel: W's friend is James Gibson, clergyman at Bishop's Waltham, who saw duty as a naval chaplain in North America; he provided W with graphic accounts of Wolfe at Quebec (1759), as well as details of natural history in that region.

39 *Andalusia . . . there*: knowledge of climate in southern Spain came to W from his brother, John, chaplain at Gibraltar 1756–72.

lumping weight: a 'lumping pennyworth' meant 'plenty for one's money', so W judges a house mouse to weigh a good ounce or more.

much: so much that on 13 Jan. 1768 NJ records they appeared 'as if scorched in the fire'.

40 *vents . . . nose*: the vents referred to here are, in fact, part of a deer's scent sacs; they have no function related to breathing.

41 *I. xi*: this is a good example of one of W's strengths. Not only is he a convincing field naturalist, but in the library also he is concerned to return to original sources.

bullfinch . . . animals: prompted by a correspondent (W's brother-in-law, Thomas Barker of Lyndon, Rutland) W referred to this

incident in a letter of 14 Sept. 1773; he observed the phenomenon about ten years previously, when (as a curate) he 'first
undertook the church of Far[r]ingdon', the bullfinch belonging
to his host where he 'used to dine on a Sunday' (*Selborne*, ed. T.
Bell (1877), ii. 101). The form of (black) melanism W reports is
a fairly common occurrence in captive finches fed a diet with an
excess of hemp seed.

42 *young . . . &c*: nidifugous (nest-fleeing) birds are both precocial,
capable of locomotion almost immediately on hatching, and
ptilopaedic, covered with down. Characteristic of many ground-
nesting species, they run, as W observes, to obtain protection
from predators and include 'waders' (plovers, sandpipers, gulls,
snipe, and so forth) as well as most game birds (partridge,
pheasant, grouse).

44 *Ray . . . Johnson*: W's authority as a field naturalist is noticeable
here. Ralph Johnson, of Brignall (near Barnard Castle) where
he was vicar and schoolmaster, was described by Ray (in his
preface to Willughby's *Ornithology* (1678)) as 'a Person of
singular skill in *Zoology*, especially the *History* of *Birds*'.

Nomina: there is a predilection in W to use pre-Linnaean
terminology, esp. that of his master, Ray; but here, writing to
his modern contemporary, Pennant, he gives Linnaean names.
Cf. the similar list in DB 1 which, although of a later putative
(and real) date gives preference to Ray. The queries attached to
two species indicate caution concerning status (not uncertainty
over identification) which seems to have been dispelled by the
following year—see DB 1.

46 *venom . . . oil*: belief in the poisonous nature of toads reaches
back to antiquity, one authority asserting the toad capable of
killing 'by its very look and breath', another claiming its flesh to
be poisonous if eaten.

migration . . . ponds: evidence for the migration of frogs was
adduced by Ray in his argument against spontaneous generation—which is what some unobserving scientists believed on
the sudden appearance of myriads of tiny migrating frogs after
rain.

47 *Merret . . . Switzerland*: W was correct; the European distribution
of this tree frog, *Hyla arborea*, is confined to central (and
southern) regions. Christopher Merret (1614–95) was the first
keeper of the library at the Royal College of Physicians and
published *Pinax Rerum Naturalium Britannicarum* (1667).

Ellis: John Ellis (?1710–76), with whom in his original letter to Pennant, of 16 June 1768, W claimed he was 'once acquainted', established his reputation with *An Essay towards a Natural History of the Corallines* (1755), in which he showed the animal status of coral.

iguana . . . frogs: in W's time the distinction between lizards, which are land reptiles, and newts, which are amphibious, had not been established. The 'lizard' in W's well (TP 20) was a male crested newt, *Triturus cristatus*; it carries 'fins' on its back and tail during the breeding season.

viper: in an introduction to his study of reptiles, Pennant wrote that the 'more fatal they are, the more deeply we should enquire into their effects, that we may be capable of relieving those who are sufferers, and secure others from the same misfortune', *BZ* (1769 edn.), iii. p. ix.

48 *visit . . . house*: W travelled to Fyfield (near Andover) on Sat. 2 July 1768 to stay with his brother Henry; he returned to Selborne on Sat. 16 July.

49 *Mazel*: Peter Mazell (fl. 1761–97) drew and engraved many specimens for Pennant's *BZ*.

curing . . . papers: extensive details of this 'cure' are given in an appendix to Pennant's *BZ* (1769 edn.), iii.

50 *notes*: the distinctions made by W still stand, and are essential for easy identification in the field. Assembling the data provided here and in DB 1, we have:

black legs: chiffchaff, *Phylloscopus collybita* ('Smallest willow-wren . . . chirps');

flesh legs: willow warbler, *P. trochilus* ('Middle willow-wren . . . sweet plaintive note');

flesh legs: wood warbler, *P. sibilatrix* ('Largest willow-wren', 'sibilous grasshopper-like noise').

See also TP 16 (para. 2), but note that the 'grasshopper-lark' of the following para. in TP 16 is the grasshopper warbler, *Locustella naevia*, and not *Alauda trivialis*, tree pipit, of the subsequent table.

52 *gentleman from London*: the *Selborne* MS identified this person as 'my Brother from Fleetstreet'.

ouzels . . . counties: that ring ouzels are passage migrants in southern England was first recorded by W. His observations provided Pennant with a contradiction since his informants in Scotland claimed the birds as resident. Confirmation of the bird's winter quarters (southern France, Iberian peninsula, north-west Africa) was not established until well into the next century.

53   *examination . . . creation*: see TP 14; the spiracula are musk glands. Pennant's anatomist was one of the Hunter brothers, William or John, both of whom were active in London and interested in quadrupeds.

  lately: from the absence of numerical readings and the nature of observations recorded we know from NJ that W left Selborne on Thurs. 29 Sept. 1768 and returned on Sat. 1 Oct. During the visit to Chilgrove (near Chichester) he noted flocks of starlings, but nothing similar for stone-curlews. His host at Chilgrove, John Woods, was both a relative, father-in-law to Rebecca (the elder of W's two sisters), and a reliable informant about natural history. From the beginning of the next letter (TP 21), in which W expresses delight that Woods has bought a copy of Barrington's NJ, it can be inferred W took his own NJ to Chilgrove to share the purposes of such a record.

54   *propensity . . . deceived*: an expression of W's time, born in the mistrust engendered by the Civil War of the preceding century and the subsequent religious controversies. The passage also echoes W's belief in a rational world that could be fully understood only if sufficient intelligence and observation were brought to bear on its mysteries: superstition, he thought, was unworthy of educated people.

55   *number . . . view*: in part because of the less wooded and less hilly terrain of these counties. W knew fenland and its adjacent areas from a long visit to the Isle of Ely (and Lincolnshire) in 1746.

57   *British Zoology*: this was the first edition (1766).

  tea: and not only tea, since the hermitage (a wooden, thatched structure on Selborne Hanger) was also the scene of melon-feasts (see *GW* 31–2).

  Indian-grass: purple moor-grass, *Molinia caerulea*, used by Hebridean fishermen to attach hooks to nets and lines.

58   *this year*: 1768 was so wet at Selborne, from early June to December, that farmers had to 'sow wheat again briskly' as late as 10 Dec. (NJ).

  Guernsey lizard . . . garden: W's 'specifically the same' (i.e. of the same species) is unlikely. The 'green lizard' of TP 22 (para. 3) and of TP 17 (final para.) will have been male sand lizards in their breeding colour. Guernsey has only one lizard, the true green lizard, *Lacerta viridis*, and those at Pembroke will have been brought direct from the island, with which Pembroke had close associations.

58 *trees*: oaks, cut down at the time of the destruction of the Hall (the seat of the Heron family) by fire in 1791.

last Michaelmas-day: this was the occasion of W's journey to Goodwood (see TP 28) and Chilgrove (see TP 20, note on 'lately'). NJ for 29 Sept. 1768 records: 'Swallows cluster on the bushes in the burnet', the likely site for this observation being the common land along the northern edge of the R. Rother, west of Rogate. Pennant paraphrased and incorporated the whole account in his 1770 (fourth, and supplementary) volume of *BZ*, giving due acknowledgement to his 'ingenious [clever] correspondent, the Rev. Mr White'.

60 *Mr Banks . . . sea coast*: Banks, whom W met twice in London (May 1767 and May 1768) was correct, as this cockchafer, *Melolontha fullo*, is occasionally found on the Kent coast.

migration . . . counties: before W informed him otherwise, Pennant also believed ring ouzels to be restricted to Wales and Scotland. He included W's observations in *BZ* (vol. iv, 1770) alongside continued belief that the birds were resident in Britain's mountainous regions.

61 *My bird . . . inquiry*: the 'new *salicaria*' of this para. is a sedge warbler, *Acrocephalus schoenobaenus*. W's care to confirm the information that the bird sang all night (see TP 25, para. 2) is typical of his procedures.

Ingenious men . . . Atlantic: in the real letter of this date to Pennant, W referred to Alexander Catcott (1725–79), who had included in the second edition (1768) of his *Treatise on the Deluge* an essay, 'The Time when, and the Manner how, America was first Peopled'.

63 *classic*: shortly before the date of this letter W agreed to superintend the education of his nephew, 'Gibraltar' Jack, who was to arrive at Selborne in Sept. 1769. At this point he would already have been considering his future responsibilities, since we know uncle and nephew were to read together some of the classic Latin authors.

64 *Edwards's drawing*: one of many by George Edwards for Hans
    Sloane, whose natural history specimens and drawings formed,
    on his death in 1753, the nucleus of the British Museum's
    collection in this field. The bird, a woodchat shrike, *Lanius
    senator*, is a rare visitor to Britain, wintering in north-east Africa
    and breeding in the Iberian peninsula, France, and east to
    Poland. W's specimen was sent by his brother, John, from
    Gibraltar in the first 'box of curiosities' to arrive from there at
    Selborne earlier in August.

65 *you spent ... highlands*: Pennant was in the Highlands for two
    months. He was so pleased with his discoveries, and with the
    people, that he returned in 1772 with a botanist (John Light-
    foot) for a longer tour, most of which he spent exploring the
    Hebrides.

66 *work*: this is vol. iv of *BZ* (1770). Pennant included in it many of
    the creatures he had seen during his tour, including several that
    W comments on in this letter—e.g. the snow bunting, *Plectro-
    phenax nivalis* (W's 'greater brambling, or snow-fleck'), alpine
    hare, eagle owl, sedge warbler (W's 'fen-*salicaria*').

69 *bat-fowlers*: whilst 'larkers' dragged nets at dusk and dawn, 'bat-
    fowlers' worked in the dark. Hay, straw or similar materials
    were attached to poles, lit, and brandished near the hedges:
    birds flew towards the lights and were then struck down with
    the 'bats', long staves with bushy tops.

70 *it was hoped ... breed*: a scientific approach to breeding began in
    the middle of the eighteenth century and many experiments
    took place in the 1760s and 1770s. As well as a desire to improve
    stock, a belief was current that fertile hybrids might lead to the
    creation of new species.

71 *American moose ... creature*: interest in this comparison was a
    matter of keen debate since the American moose was known to
    be larger than what was taken to be its European form (elk).
    One position, accepted by Pennant (see his *History of Quadrupeds*
    (1781), i. 95–8), believed moose to be of greater size than elk
    because America provided 'larger forests to range in and more
    luxuriant food', but another position acknowledged the two
    creatures as different species: both arguments were a response
    to a general (Eurocentric) belief that creatures at the periphery
    of the known world were less developed than those nearer the
    centre.

72 *male otter ... shot*: from NJ we know this otter was killed in Oct.

1784. Its inclusion in a letter with a date of 1770 is, therefore, somewhat startling—but from it we can learn much about W's editing of his real letters and the shifting criteria he adopted, almost letter by letter, in the preparation of *Selborne*.

73 *My friend . . . dimensions*: W's friend was James Gibson, a naval chaplain, who saw service in North America. As well as giving W details about moose, he had sent him in 1759 a graphic account of Wolfe's success at Quebec.

75 *improbable facts . . . affection*: W was, in a sense, right as the woodcock has been observed carrying its young, albeit between its thighs.

76 *hirundo rupestris . . . matter*: in his real letter to Pennant of 12 May 1770 W had (for the first time) revealed that on occasion he received 'communications in the natural way' from a brother at Gibraltar. It was one of these 'communications', a crag martin (*Hirundo rupestris*), that provoked particular interest and prompted W to seek assistance from Pennant.

summer-birds of passage: for details of these, see DB 7 (para. 11).

79 *This fly . . . Linnaeus*: now *Piopila casei*, a species of cheese-fly. The 'harvest bug' of the preceding para. is the larva of a mite, *Trombicula autumnalis*, and the 'turnip-fly' of the following para. is a flea-beetle, one of the *Chrysomelidae*.

modern entomologists . . . f.4: W had considerable respect for the work of Étienne-Louis Geoffroy (1725–1810), since he sent the *Histoire . . . des insectes* (2 vols., 1762; 2nd edn. 1764) to John at Gibraltar in 1770, saying that the plates of insects were 'the best' he had ever seen (P. Foster, 'The Gibraltar Correspondence of Gilbert White', *Notes and Queries*, 32 (June 1985), 320, 323). The 'star-tailed maggot' of this para. is the larva of a dipterous fly, *Stratiomys chamaeleon* (one of over fifty British species of soldier-fly which frequent aquatic plants), whereas the 'curvicauda of old Moufet' (Thomas Moffet (1553–1604), whose *Theatrum Insectorum* was published posthumously in 1634) is the horse bot, *Gasterophilus equi*, the larvae of which are licked off horses' legs, pass through the intestines, and subsequently pupate in the ground. W's reference to Geoffroy is elliptical—it must be read as vol. [t(ome)] ii, plate 17, f[igure] 4.

destroying them: including, it was hoped, natural means 'as we do cats against mice' (W. Curtis, *Fundamenta Entomologiae: or, An Introduction to the Knowledge of Insects—Being a Translation [from] Linnaeus* (1772), 10.

81   *orange*: a good illustration of W's recourse to homely comparisons
     to convey ideas of size. The hair-ball reported here was noted in
     NJ for 1 Mar. 1770 and, together with the account of the
     peacock display, formed part of a real letter to Pennant of 19
     July 1771.

     species of bat . . . sex: both specimens, obtained on successive
     evenings (see NJ 7 and 8 Aug. 1771), were noctule bats, *Nyctalus
     noctula*, W being the first to describe this species in English.

83   *migrations*: all these movements are entered in NJ for Aug. and
     Sept. 1771.

     martins . . . fledged: this was at Lasham, where W stayed overnight
     on 1 Oct. 1772 on a journey from Selborne to Oxford.

     more northern . . . say: claimed thus ('SWALLOW *goes under water*')
     for 17 Sept. 1755 at Uppsala, Sweden, in 'The Calendar of
     Flora' by Alexand. M. Berger, first published in English by
     Benjamin Stillingfleet and included in his *Misc. Tracts* (2nd edn.,
     1762).

84   *new edition of the British Zoology*: known as the 4th edition, it was
     published 1776–7, and incorporated many of W's observations.
     To assist Pennant in the real letter (and in *Selborne*), W provided
     for each of the following paragraphs (except those devoted to
     the 'house-swallow' and to 'Wagtails') a footnote reference to
     BZ (vols. i and ii, 1768).

85   *Titlarks*: tree pipit, *Anthus trivialis*, not meadow pipit, *A. pratensis*,
     which sings in its ascent as well as descent and which W
     sometimes refers to as 'small field lark'.

     Whin-chats . . . year: whinchats, *Saxicola rubetra*, are strictly
     summer migrants, but can be mistaken for juvenile and female
     stonechats, *S. torquata*, in winter plumage.

86   *swift*: at this point in his comments on *BZ*, W moves his
     attention to vol. iv (1770). Pennant's description compares the
     swift with the 'fabulous history of the Manucodiata, or bird of
     Paradise [which] was believed to have no feet, to live upon
     caelestial dew, to float perpetually on the *Indian* air, and to
     perform all its functions in that element' (p. 15).

87   *eels . . . mysterious*: that eels breed only in the Sargasso Sea was
     not known until much later; the 'threads' are parasitic worms.

88   *marsh titmouse*: differentiation of this tit from the willow tit was
     not made for almost another century.

89 *mistake*: a good example of W's integrity in relation to the behaviour of creatures.

90 *how those species ... months*: this entire letter was part of a real letter to Barrington, dated 15 Jan. 1770. W includes it here since the content follows from that of TP 39 and TP 40 and because it provided information relevant to Pennant's enquiries.

92 *wilderness*: a biblical echo of the provision of quails and manna (the excretion of certain scale insects) to the Israelites: see Exodus 16: 13–15.

93 *faunist ... received*: this whole para. provides an instructive insight into the responsibilities of a 'faunist'. Natural history in W's time embraced more fields of study than it does today and attempted a comprehensive account of a geographical region; the sole volume bearing any resemblance to W's expectations was John Rutty, *A Natural History of County Dublin* (1773).

Fort William ... omitted: Fort William was constructed in the reign of William and Mary, but the other forts in the chain (notably Fort Augustus and Fort George) were built along the line of the Great Glen in the early Georgian period as a defence against the sympathizers of the Jacobite cause. The 'military roads' that provided rapid access to the forts were begun by Gen. Wade in 1723.

94 *honey-buzzards ... time*: NJ notes on 27 June 1781, 'The honey-buzzard sits hard', and the full details are given on an inter-leaved sheet between Sat. 23 June and Sun. 24 June 1781. A further sheet, interleaved between Sat. 7 July and Sun. 8 July 1781 (it was W's practice at this period to have NJ interleaved in this way and bound before purchase), includes the account of the sparrowhawks given in the last para. of the present letter: the continued depredations of these hawks were entered in NJ for 17 July the same year (1781).

96 *domestication ... breed*: this by-product of domestication contrib-uted to the eighteenth-century belief in the advantages of human (European) association.

cote: this dovecote in Caernarvonshire (Gwynedd) was at Glod-daeth Hall, Llandudno, below the Orme's Head. It was erected *c.* 1600 by the Mostyn family, neighbours of Pennant in Flint-shire (Clwyd), and, as Lord Mostyn kindly informs me, is still in good order.

101 *last month*: W was in London 25 Apr.–13 May 1769.

order in which they appear: W's interest in the sequence of migration was persuasive, and represented, in part, a wish to contribute to the search for a natural calendar (see Introduction).

102 *numerically*: in accordance with the numbering of W's sequence (given above) of the birds' arrival at Selborne.

104 *Nov. 2*: the real letter is dated 9 Nov.

106 *1770*: this date, 2 Jan. 1770, was added by W to his list after the writing of the original letter.

107 *boobies*: sooty terns, *Sterna fuscata*. A seventeenth-century traveller visited Ascension Island in June 1656 and recorded: 'Of sea fowle . . . are a numberlesse number; some of them . . . would lightt on our ships yards and suffer themselves to bee taken by hand like boobees' (*Travels of Peter Mundy*, vol. v, Hakluyt Society, 1936).

108 *Shakespeare*: most editions print: 'And turn his merry note | Unto the sweet bird's throat' (Act II). W obtained 'tune' from Rowe (1706) but provided 'wild' for 'sweet' himself.

109 *hundreds . . . meadows*: NJ gives for 13 July 1769: 'Vast flocks of young wagtails on the banks of the charwel'.

110 *monstrous outrage . . . belief*: even though (W implies) such creatures might have had little opportunity to profit from European values. See, above, note to 'American moose' (p. 71).

111 *Query*: this whole letter formed part of the real 15 Jan. 1770 letter; Barrington was much exercised by the life history of cuckoos and included an essay, 'On the prevailing Notions with regard to the Cuckow', in his *Miscellanies* (1781).

112 *suspicion*: whereas today explanation for species' behaviour and difference is sought at the genetic level, in W's day answers were sought in anatomical variation. In this instance Barrington tested his supposition by seeking advice from the anatomist John Hunter, who showed that in male birds (more noted than females for singing) the muscles of the larynx are much stronger than they are in hens, and that in the nightingale (the best songster) the muscles are more developed than in any other bird of equivalent size.

before: W refers to having sent Barrington a specimen of this species early in February 1770.

113 *Investigations . . . Italy*: published in *Phil. Trans.* 58 (1768).

113   *The severity . . . time*: editors sometimes criticize W for including
      the substance and phrasing of this paragraph twice—both here
      and in TP 29 (para. 1). However, the dates of both letters are the
      dates of real letters, in both of which W wrote what he printed—
      but then, it is entirely natural (and not unusual) to write about
      the same phenomenon to more than one correspondent.

115   *Kuckalm . . . Jamaica*: Tesser Samuel Kuckkahn (also Kuckhan,
      but not 'Kuckalm', which W obtained from misreading Barring-
      ton) was elected FRS 4 June 1772 as a foreign member (North
      America). His particular knowledge and skill lay in the preser-
      vation of creatures, especially birds, about which he addressed
      four letters to the Royal Society when he was in London in 1770
      (see *Phil. Trans.* 60 (1770), 302–20); he also made several gifts
      of specimens, the citation for 1772 reading 'Specimens . . . from
      Jamaica'.

      W's interest in foreign fauna had been launched principally
      by the specimens he received from his brother at Gibraltar, and
      it is highly appropriate that this letter (devoted mainly to
      extraparochial fauna) should be written out of the parish, at
      Ringmer where he stayed with his aunt, Rebecca Snooke, from
      4 Oct. to 19 Oct. 1770.

116   *titlark*: meadow pipit, a common foster parent for the cuckoo. W
      records the whole incident in NJ at 25 July 1770.

      cuckoos . . . prey: because of its barred plumage and the disap-
      pearance of the host's eggs, traditional writers classed the bird
      with other predators, such as crows and woodpeckers. W had
      made his observations at Oakhanger, within Selborne parish.

118   Sir: most previous edns of *Selborne* begin this letter with a single
      sentence: 'The birds that I took for *aberdavines* were reed-
      sparrows (*passeres torquati*).' Its inclusion was an error on the
      part of the compositor: for discussion, see *GW* 200.

      see . . . cocks: males leave the flocks of mixed sexes in late winter
      to establish breeding territories.

119   *wheat-sowing . . . oats*: from September (wheat-sowing) to March
      (barley and oats).

      woodcocks . . . experienced: this evidence was adduced by Barring-
      ton as part of an argument against migration—see below, note
      to 'bird may travel . . . Gibraltar' (p. 122).

120   *breeding*: woodcocks are resident and migrant breeders, as well
      as passage and winter visitors—and may well have been in W's

day. The birds are silent when flushed, feed at dusk, spend the day at rest, and seek seclusion at breeding time, characteristics enough (in a period of chance observation and record) to suggest numbers were fewer than they actually were.

There fell . . . half: information obtained from Thomas Barker, W's brother-in-law, who reported his meteorological observations regularly not only to W but also to the Royal Society.

121 *Belon*: Pierre Belon (1517–64), French traveller and naturalist, published *L'Histoire de la nature des oyseaux* (1555).

122 *bird may travel . . . Gibraltar*: Barrington accepted this view. What his argument advanced in 'On the periodical Appearing and Disappearing of certain BIRDS, at different Times of the Year' (first publ. in *Phil. Trans.* and included in his *Miscellanies* (1781), was doubt concerning a narrow definition of migration, that is, 'a periodical passage by a whole species of birds across a considerable extent of sea' (by which he meant the North Sea or the Atlantic).

123 *skylarks . . . also*: larks dust and also wash, although the latter is carried out only in the rain, not in standing water.

Andalusian birds: several of the Gibraltar cargoes arrived in London to be overseen by W's brother Thomas, who arranged appropriate assistance for identification of contents. Marmaduke Tunstall (1743–90) was a bird collector with a museum in Welbeck St., his collection (some specimens from which still survive) forming much later the nucleus of the Hancock Museum at Newcastle-upon-Tyne.

Tenant: not identified positively, but possibly Andrew Ten[n]ant, bookseller of Bath (d. 1788), whom Barrington may have met at the time he was Recorder at Bristol (from 1764) or, possibly, through Benjamin White (W's brother who specialized in natural history books). Fieldfares, although now known to nest in northern England and Scotland, were winter visitors only in W's time—which is why Tenant's account is questioned.

124 *musical friend . . . B flat*: W's 'friend' was his brother, Henry, with whom he stayed at Fyfield (near Andover) from 5 Feb. to 22 Feb. 1771. The topic was prompted by Barrington, who was preparing material for 'Experiments and Observations on the Singing of Birds', *Phil. Trans.* 63 (1773). Work in the 1980s has shown that sonagrams enable individual tawny owls to be identified and (to change the creature) that the buzz of the bumble bee, *Bombus terrestris*, is normally around D below middle C.

124 *pitch-pipe*: measurements of eighteenth-century pipes give a pitch
     of 425 KHz—much lower than that in use today.

125 *you observe*: and not only in correspondence since in an essay,
     'On the Linnaean System' (*Miscellanies*, 1781), Barrington was
     very cautious in his approval and, for the field naturalist,
     strongly advocated the continued use of Ray's descriptive
     names.

126 *Foreign . . . terms*: it is here that W demonstrates both a strength
     and a weakness. His opinion of 'foreign systematics' had been
     hard-won during the identification of *Hirundo hyberna*, crag
     martin (see TP 32, and correspondence cited above (p. xvii)
     from William Sheffield).

127 *as these animals . . . fail*: modern field studies suggest the contrary
     hypothesis—that winter flocking lessens the danger of indi-
     viduals going without food for long periods.

128 *us*: W's companion was his brother Thomas, who had accom-
     panied him from Selborne to Ringmer on 1 Nov. 1771. The
     whole account of the observations, much in the terms of the
     present (putative) letter to Barrington, formed part of W's 8
     Nov. 1771 letter, from Ringmer, to his brother John at Gibraltar.

     formerly . . . writing to you: W means here his letter of 8 Oct. 1770
     that formed the basis of DB 7.

131 *conversation*: as far as extant records show, Barrington never
     visited Selborne and all W's meetings with him took place in
     London, especially in the period 1770–5.

132 *preposterous*: from Lat. *prae*, 'before', and *posterus*, 'after'; hence, a
     murder which made 'that first which ought to be last' (John-
     son). Compare with W's use of the word (albeit a century earlier
     and in relation to the act of a father):

> Death from a father's hand, from whom I first
> Receiv'd a being! 'tis a preposterous gift,
> An act at which inverted nature starts,
> And blushes to behold herself so cruel.
>   (Denham, 1615–69)

134 *owl . . . water*: birds do not sweat, so the main purpose of
     drinking is to replace fluid lost through excretion and through
     breathing. The heavy water-takers, therefore, are the seed-
     eating species, birds with a succulent diet (fruit, nectar, insects,
     or—as with owls—flesh and blood) rarely needing to drink.

It will be . . . appearance: this paragraph of explanation for the inclusion in *Selborne* of papers previously published in *Phil. Trans.* provides a useful reminder of the variety of editing processes present in the text. The four cited letters exist in three forms: first, as real letters to Barrington; second, as letters read at the Royal Society (Letter 16 on 10 Feb. 1774, Letters 18, 20, and 21 on 16 Mar. 1775); third, as letters in *Selborne*—which include the addition of some observations made later than the dates of the real letters to Barrington.

135 *dipterous insects . . . feathers*: a large number of different creatures are parasitic on birds, many of them being adapted to the varied ecological niches present on the host. Here W refers to louse-flies, some forty of which have been reported on a single house martin.

entomologist: René-Antoine Réaumur (1683–1757), whose thermometer scale is still used. His work on insects was introduced to W by his brother John (but not until 1776), and in *Selborne* W's textual reference is to Plate 11 of vol. iv.

138 *close air of London*: the air in London was so poor in quality (from the smoky coal used) that visitors from abroad complained even in summer that it 'must necessarily have an effect upon the constitution of the inhabitants' and even attributed to it a main cause of the 'English Melancholy' (Pierre Jean Grosley, *A Tour to London*, 2 vols. (1772)).

140 *society . . . them*: fellows and guests of the Royal Society, fellows alone being permitted to introduce and read papers from non-fellows.

ravished . . . satisfaction: for W pleasure in landscape rested in the variety offered. Travellers 'on the downs of Sussex', as Ray wrote, 'enjoy'd that ravishing Prospect of the Sea on one Hand, and the Country far and wide on the other'.

plastic power: this phrase reveals one strand of W's philosophical inheritance. The Cambridge Platonists (of the later decades of the preceding century) believed growth, in vegetation and creatures, to be sustained by a 'plastic' power. In describing chalk hills as he does W extends this power to the formation of the inanimate world—knowledge of chalk deriving from marine animals not then being established.

141 *this diversity . . . downs*: W noted this distinction (26 Sept. 1769 in NJ) at the time of an earlier visit to Ringmer. Cross-breeding

was only just beginning and geographical areas were restricted to a local breed. W's black-faced, poll (hornless) sheep in the east were Southdowns; those in the west, where the ground was wetter and richer, were Dorsets. For the flocks of Laban and Jacob, see Genesis 30.

143 *rookery . . . mild*: W's interest in rookeries was strong, and he regularly recorded in NJ behaviour at a rookery he passed almost weekly from 1761 to 1784 on his journey from Selborne to Farringdon where he conducted Sunday service.

Shrove Tuesday . . . February: this was before the revision of the calendar in 1752, so eleven days need to be added to W's dating.

148 *another bird . . . eggs*: the source for this anecdote was W's brother, John, who obtained it from a Lancashire neighbour, Sir Ashton Lever, the founder of the Leverian Museum in London.

149 *pamphlets . . . seasons*: the winter of 1773–4 led to considerable unrest because of the high prices for corn. Many groups of men ('combinations'), desperate to feed their starving families, caused much disturbance.

150 *hirundo esculenta*: a term that embraced several species of so-called edible (bird's nest) swiftlets. However, the nests of *Collocalia esculenta*, glossy swiftlet, contain too much extraneous matter for even the hungry, and bird's nest soup relies on nests from *Aerodramus* sp., esp. *A. fuciphagus*, edible-nest swiftlet. White's information came from Mathurin-Jacques Brisson (1723–1806), a French chemist and naturalist, who published *Ornithologia* (1760).

fera natura: a wild creature, not associating with human habitation. An underlying assumption here is that in Eden all creatures welcomed human company, but that since the time of the Fall and of the later Flood (when all creatures lived peaceably in the ark) some species, formerly domesticated (tame), have become wild (feral).

152 *pulex irritans*: not the human flea, but *Ceratophyllus styx*, one of 125 species of bird flea, many of which are species specific; up to 2,000 fleas have been recorded in a single nest.

153 *skirts of London . . . neighbourhood*: several of W's relatives lived north and south of the Thames and he himself often crossed 'Saint George's-Fields' (now the general area of Waterloo Station) as he moved between South Lambeth, the home of his brother Benjamin, and the City, where his brother Thomas

lived. However, this observation was entered in NJ at 30/31
Mar. 1776 and was an addition to the original letter.

157 *fed on the wing ... dams*: an opinion W was to revise, for NJ
records on 20 July 1784: 'Saw an old swift feed [its] young in
the air: a circumstance which I could never discover before'.

I untiled ... build: NJ shows W to be at Fyfield at this time and
records for 30 June 1774: 'I procured a bricklayer to open the
tiles in several places round the eaves of my Bro:'s brewhouse,
in order to examine the state of swifts' nests'.

swifts ... together: young swifts are less demanding than young
of some other species and, in cold weather particularly (when
the supply of insect food is diminished), retard their development significantly: the nestling period may extend to eight
weeks.

Do they ... breeding: swifts moult twice—after breeding, and in
the spring.

159 *carry ... forward*: this is only when at rest or in museum
specimens; when the foot is functional it is shown to be
heterodactyl, toes one and two opposing toes three and four in a
lateral and inward movement which is ideal for enabling the
swift to grasp its soft, pliant nesting materials, most of which
are airborne.

160 *a succession ... spots*: speculation on this was included in NJ (3
June 1775), and in response Barrington wrote on W's journal
page: 'Cut off a claw of one of them caught in its nest. Observe
then whether the same bird does not return next year.' This
mode of marking (the same as that used by Edward Jenner in
his work on avian migration) seems not to have been pursued
by W.

swiftness of a meteor: relative to many species, and contrary to
popular belief, swifts fly slowly, at no more than 40 km/h—
eider duck, for instance, travel at almost twice that speed (76
km/h). The fastest known avian movement is the stoop of a
peregrine which has been measured at 180 km/h.

161 *Sept. 13, 1774*: the last paragraph of the preceding letter postdates
this letter—correctly so, since it was an addition to the original
letter; see above, note to 'It will be ... appearance' (p. 134).

Do these different dates ... migration: data of the kind given, from
different centres, were crucial for establishing patterns of migration. Information for Devon came from Sampson Newbery (of

Exeter College, Oxford), and for Lancashire from W's brother
John, now returned from Gibraltar.

161 *A farmer . . . dung*: W's comments in *Selborne* on husbandry are
few, but from his garden records we know he kept an annual
note of dung bought for his garden, and regularly entered in NJ
details about his own cropping of meadow grass for hay. The
farmer at Weyhill was 'Farmer Cannings', W obtaining the
details given in the letter on 7 June 1774 as he returned from a
visit to his brother at Fyfield.

162 *The missel-thrush . . . legumens*: these observations of the missel-
thrush were entered in NJ for 23 Apr. 1774; interest in the
Welsh name derived from Pennant, who included it in his
description of the bird in *BZ* (vol. i, 1768).

164 *a gentleman . . . veneration*: a warm tribute to W's father.

why these apterous insects . . . excursion: and, similarly, why W chose
this date, 8 June 1775, to write about an experience of 30 and
more years past. There are, however, some extant clues: late in
1774 and early in 1775 he was writing to a nephew, Samuel
Barker, about the beauties of literature. This led to inspection
of a newly-published glossary to Chaucer and from this he
entered in NJ (28 Jan. 1775) a note that Chaucer associated
gossamer with thunder, with the tides, and with mist, as
unexplained mysteries in the world. Alongside this literary
source (or, perhaps, prompted by it), W sent in Mar. 1775 an
account of gossamer to his brother John, at Blackburn, as a
contribution to a comparison between the climates of England
and Gibraltar.

165 *spirit of sociality . . . instance*: this entire letter, as evidence of
association across species (even to the point of incongruity), was
included by Barrington in his essay on the cuckoo; see above,
note to '*Query*' (p. 111).

167 *these vagrants . . . discovered*: work on language has led modern
authorities to trace gypsies back to India; they are thought to
have left there well before AD 1000, and to have entered Europe
by at least the fourteenth century. The first documentary record
of presence in Britain (1505) is from Scotland.

168 *rush*: there are two species here—*Juncus effusus*, soft rush, and *J.
conglomeratus* (or, *J. subuliflorus*), now known as compact rush;
both would have been readily found in the Selborne area and
both were used for rush-lights.

Decayed: in W's time, and for more than a century before, 'decay' was a natural consequence of life. Usually applied to old age and advancing decrepitude, here it embraces whole families and could be read today as equivalent to 'unemployed'.

169 *a poor family . . . round*: the Bank of England conversion scale suggests a factor of about 60 in order to relate W's costs to those of today: thus a year's lighting could be provided for about £14.00.

170 *the very poor . . . economists*: a view shared by William Cobbett fifty years later in *Cottage Economy*, where he advised the use of rush-lights. 'Economy', as Cobbett stressed, is to do, not with parsimony but with good management.

an idiot-boy . . .: at first sight this letter seems out of place in a natural history, but in W's century natural history or, as Robert Plot put it in his *Oxfordshire* (2nd edn., 1705), the '*History of Nature*' included 'the unusual Accidents that . . . attended' men and women through life's 'extravagancies and defects'.

171 *stools*: used to raise the skeps (hives) above the ground to hinder depredations from mice—rather like staddle-stones under a granary.

Thou . . . : quotation untraced, but the Wildman referred to must be either Thomas Wildman, who published *A Treatise on the Management of Bees* (1768) or his nephew Daniel Wildman, who also wrote and who established contemporary fame by trick performances with bees.

172 *cut down . . . together*: this is a fair example of W's humanity. Superstitious belief and practice were anathema to him, and we must conclude that the tree that 'continued to gape' was removed, at least in part, because of its reminder of an ineffectual (and barbarous) 'cure'.

173 *The late . . . it*: a version of this line occurs in 'Baucis and Philemon', an imitation of Ovid by Swift. As is common in quotations that White cites, the relevance is only partial, Swift's tree being not an ash but a yew.

174 *In Newton-lane . . . dusty*: the detail of this observation, together with some speculative comment, occurs in NJ at 7/8 Oct. 1775.

175 *the small . . . affected*: recorded in NJ for 27 May 1775, the real interest lying in W's noting the phenomenon after ten or so weeks of drought. The natural historian is interested both in the

usual, the expected, the routine, but also in the unusual, the
extreme: from these last much valuable speculation ensues, and
W sought advice from several authorities, including his brother
Thomas.

176 *Hence . . . resource*: since the earth cools more rapidly than water,
the supply for upland ponds is derived mainly from the surrounding ground rather than from direct precipitation on the
surface of the water.

177 *Not long . . . swallowing*: the substance and phrasing of this para.
occur in a letter of 12 Aug. 1775 to W's brother John at
Blackburn.

178 *we surprised . . . sun*: this was at Ringmer, where W and his
nephew, Samuel Barker, were staying with W's aunt, Rebecca
Snooke. The capture and investigation of the viper are entered
in NJ for 3 Aug. 1775, and were included in the letter of 12 Aug.
(referred to above) sent to John; the letter was written at
Ringmer.

180 *he had a boar . . . off*: a similar practice is adopted today on the
rhinoceros—to prevent poachers killing for ivory. No detrimen-
tal effect on potency has been observed, although opinion on
such a matter is not unanimous. Mr Edward Lisle was a
Hampshire man, from Crux-Easton, south of Newbury; he died
in 1722 but his *Observations in Husbandry* was published posthumously in 1757.

181 *two litters in the year . . . quadruped*: there appears to be no extant
correspondence about this topic, or indeed that of the preceding
letter. NJ, however, records on 4 Mar. 1775 the age of potency
of pigs and sheep (within less than a year of birth) as compared
with horses and cattle, and it is evident that at this stage in the
composition of *Selborne* W was working through NJ to write up
topics as they occurred.

182 *why so cruel . . . offspring*: W was much moved by this incident
and referred to it again as late as 18 Apr. 1790 (NJ) when giving
an account of a cat suckling three young squirrels. The epigraph
to the letter, however, is somewhat misleading. Although it
refers to cross-species suckling (of a man by a tigress), the
meaning is solely metaphorical: the speaker, Dido, berates
Aeneas for his heartless departure from Carthage and taunts
him with having been brought up by tigers.

183 *are out . . . winter*: NJ records such an occurrence on 3 Dec. 1776,
with the weather as 'still & warm'.

184 *neighbourhood*: this was at Kingsley, a parish adjacent to Selborne to the north-east. Many of the details for these dates (26, 27 Mar. 1777) reached W by report as he was in London from 9 Mar. to 9 Apr.

185 *torpid insects . . . heat*: the absence of aerial insects (again, due to the cold) is a contributory factor. Limited periods of avian torpidity and nocturnal hypothermia (*controlled* reduction of metabolism) have been documented this century for swifts, nightjars, and humming-birds.

187 *One cause . . . touch*: diet is not implicated as a cause of leprosy.

Potatoes . . . reign: potatoes had a bad press in southern England in the reign of George II for they were associated with Popery, leprosy, venery. W himself introduced potatoes to Selborne in the mid-1750s and paid villagers (gave them 'premiums') to grow them. Such was his success that by 1787 NJ records early in Oct.: 'The quantity of potatoes planted in this parish [is] very great, & the produce . . . prodigious.'

188 *Lord Cobham . . . Beaconsfield*: W's choice of gardening celebrities is somewhat selective, but Edmund Waller (1607–87), Member of Parliament both before and after the Commonwealth, as well as developing his estate (Hall Barn), publicized garden improvements in fashionable poetry; Lord Ila (Islay), Archibald Campbell (1682–1761), 3rd Duke of Argyll, was noted for his citrus fruits, which he grew at Whitton, Hounslow, on an unheated wall protected in winter by glass lights; and Lord Cobham, Sir Richard Temple (?1669–1749), rebuilt Stowe and laid out gardens claimed by contemporaries as the acme of taste—W visiting there on several occasions (see *GW* 167).

191 *Anathoth*: a Hebrew word meaning 'echoes' and the name of a biblical city.

192 *should any . . . expense*: later (see DB 60) one of W's nephews was to explore Selborne's echoes with gunfire, but W's own improvements were less 'loquacious'—an arbour, ha-ha, hermitage, and so forth (see *GW* 207–8).

193 *Lib*. IV: W's quotation is from *De natura rerum* (Of the Nature of Things) by the Roman philosopher Lucretius (99–55 BC). This work, in verse, is a lyrical exposition of the Greek philosophy of Epicurus that the sole basis of knowledge is sense perception: such a view gives no room for belief in superstition and was, therefore, much in accord with W's own views.

194 *the numbers returning . . . retiring*: compensation for the well-documented depredation of swallow and martin populations during breeding and migration is achieved by the number and size of broods; swifts, nesting once a year and, normally, laying only two eggs, are a much more stable species.

196 *two blades . . . before*: quoted from Swift's King of Brobdingnag in *Gulliver's Travels* (1726), and a sentiment endorsed by Stillingfleet in 'Observations on Grasses'. This last was included in *Misc. Tracts* (2nd edn., 1762) alongside Stillingfleet's advocacy of a search for a natural calendar—which was also designed as a contribution to the reformation of practices in husbandry.

 a short list . . . unentertaining: W's previous letter has stressed the useful but here, in turning to plants of little value to the husbandman, he reveals much of his own interest. This last falls into several categories: for instance, with the hellebores he is much taken by the herbaceous character of one variety; with the *Vaccinium*, by one variety growing in bog, one on dry hillside; and with the crocus (see end of letter) by the different seasons of flowering.

197 *Hudsoni*: William Hudson (?1730–93) whose *Flora Anglica* (1762, rev. edn. 1778), prepared under the auspices of Benjamin Stillingfleet, was the first British flora organized wholly on Linnaean principles. W bought a copy in 1765; it became the main stimulus for his preparation of 'Flora Selborniensis' (compiled 1766; not published until 1911), and he marked up 439 species as pertaining to the Selborne flora. *Blackstonia perfoliata*, yellow-wort, in the naming of which Hudson commemorated a fellow London apothecary (John Blackstone, d. 1753), remains a scarce plant although locally widespread on chalk and limestone turf.

 tooth-wort: a plant W discussed in 1767 with Joseph Banks, to whom he promised to send a specimen the following spring. W appeared not to know of the plant's parasitic habit since he tried to transplant specimens into his own garden.

198 *truffles*: several records exist in NJ for truffles, but a good number of them relate to findings not at Selborne but at Fyfield, the home of W's brother Henry.

 Of all the propensities . . . blossoming: this entire para., together with its concluding couplets, is sometimes omitted from editions of *Selborne* owing to oversight; it was placed in the first edn. as an addition to DB 41 but printed *after* the 'Antiquities'.

199 *Say, what impels . . . delay*: in these few couplets W reveals his
devotion to the mysteries of the natural year. He takes 'God of
Seasons' from James Thomson, who uses the phrase in his
'Hymn' to *The Seasons*, but the glow of the 'flamy' crocus comes
from Homer's *Iliad* 14 (Pope's translation), where it accompan-
ies Jove's passion, amidst the snowy heights of Mount Ida, for
Juno, his queen of heaven.

apparent: this quotation from Virgil, *Aeneid* 1, is a good example
of W's tendency to adapt a literary source to the concerns of
natural history. In Virgil, Aeneas recognizes his immortal
mother, Venus, disguised as a simple huntress, not by her
speech but by her step. Today, what W describes would be
termed the 'jizz' of a bird, a term first deployed by T. A.
Coward, *Bird Haunts and Nature Memories* (1922).

202 *The language of birds . . . understood*: avian calls express most
aspects of social behaviour; gulls, for example, possess about ten
different calls but passerines have a vocabulary of twice that
number.

203 *historians assert*: notably Livy (59 BC–AD 17), who writes in his
History of Rome that when the Gauls besieged Rome in 386 BC
the sacred geese of Juno, by whose 'clanking' and clapping of
wings the Roman guards were alerted, prevented the Capitol
from falling to a surprise attack.

205 *outlets*: W named his own grounds as an 'outlet'.

heliotrope: a structure, commonly a sundial, from which the time
may be roughly calculated. W claimed his heliotrope to be 'J.
Carpenter's workshop', which permitted him to record in Jan.
1781 that 'on the shortest day [19 Dec. 1780] the shades of my
two old chimneys fall exactly in the middle of the great window
of that edifice at ½ hour after two P: M:' (NJ 26 Jan. 1781).

206 *I nor advise . . . debates*: the 'journey' of Marcley Hill, in Herefordshire, described in *Cyder* (1708) was taken from Richard Baker
(1568–1645), whose description of an earthquake (perhaps,
landslip) of 1571 was included in his *Chronicle of the Kings of
England* (1643).

207 *hundred of Selborne*: the hundred of Selborne embraced the parish
of Selborne itself, together with those of Farringdon, Newton
Valence, East Tisted, Empshott, and Hawkley.

between the 8th and 9th . . . leap: W was not at Selborne at the time
of the landslip for he travelled to London on 8 Mar. 1774. He

returned in late March but was incommoded by 'a great cold'
and was unable to view the scene (of more than local interest
since, on one Sunday it is claimed, near a thousand people were
visitors) until mid-April or later—at which time the oaks might
well be showing leaf.

210   *I should have been glad ... settle*: this was 'on account of their
pleasing summer sound' (*Garden Kalendar*, May 1761), W con-
ducting his investigations with his brother Thomas.

211   *turf*: this was in May 1765 also (see *Garden Kalendar*).

212   *house-cricket ... dwellings*: from NJ (22 Apr. 1780) we know W
observed house-crickets in his own kitchen; one arrived there
the previous year (see NJ, 29 Aug. 1779) and bred with
conspicuous success, attracted perhaps by the new plaster of
W's 'great parlor', which was built 1777–80.

214   *visit*: W was at Ringmer 28 May to 7 June 1771 and the event
described is recorded in NJ for 6 June.

216   *stuffed with pepper*: because of the intense interest in the second
half of the eighteenth century in creatures from other countries,
there was much experiment with different modes of preservation. W's NJ for 1770, at the time he was receiving specimens
from Gibraltar, carries on the flyleaf a recipe for a 'proper
antiseptic substance for the preservation of birds', the main
ingredients being black pepper and ginger (in equal measure),
together with 'Camphire' and cloves (in half-measures) and
traces of alum, nitre, and common salt.

flamingo: W's knowledge of flamingos derived from his brother
John at Gibraltar, from whom he received several specimens—
one of which he presented to the Royal Society (26 Mar. 1772);
it was from the same source that W knew of the presence of the
stilt plover in Andalusia (see following para., final sentence).

But four pounds ... saw: conceptually, this calculation and, hence,
conclusion, is flawed—increases in weight (cubic measure) and
length (linear measure) are comparable neither across a life
cycle, for a single creature, nor across species. Of more interest
is the fact that W's attention should be taken by an extreme, a
curiosity—for attention to the monstrous, the aberrant, was to
increase in the later years of the century.

218   *The old Sussex tortoise ... post-chaises*: W was in London at the
time 'Timothy' lost 'her' (it was established only after the death
of the tortoise that 'he' was female) mistress, Rebecca Snooke

dying at Ringmer on 8 Mar. 1780, aged 86. W travelled to Ringmer on Sun. 12 Mar; his aunt was buried on 15 Mar. and he left for Selborne on 17 Mar. via 'Uckfield, Cuckfield, Horseham, Dorking, Guildford . . . the country as far as Guildford [being] new to me' (W letter of 16 Mar. 1780 to his brother Benjamin).

afternoon: 22 Apr. 1780, the day after W began this letter, as is confirmed by NJ.

219 *More Particulars . . . History*: these 'More Particulars' were printed in *Selborne* (1789) on pp. 427–8 *after* the 'Antiquities'. W's quotation from Alexander Pope is from the 'Imitation of the Second Epistle of the Second Book of Horace'; the 'scald(s) with safety' of the following para. is from Shakespeare, *2 Henry IV*, Act IV.

220 *your Miscellanies . . . deserve*: Barrington's *Miscellanies* (1781) include tracts and essays on a variety of topics—on the possibility of reaching the North Pole, on the history of the turkey, on the infant Mozart, for example; W is generously acknowledged in essays on migration, on swallows, on the cuckoo, and on the Linnaean system.

221 *One swift . . . September*: a note in NJ for 1 Sept. 1782 explains this late postscript, 'The swifts left Lyndon . . . for the most part, about Augst 23 . . . Some continued 'till Augst 29: & one 'till Septem^r 3rd!! In all our observation Mr Barker & I never saw or heard of a swift in Septem^r tho' we have remarked them for more than 40 years.'

222 *pest*: a beetle, *Pulvinaria vitis*.

Mr Lightfoot . . . Dorsetshire: NJ (May 1785) includes the substance of this entire letter, together with the detail that the Weymouth infestation was in the garden of 'M^r Le Coq'. W's correspondence with John Lightfoot (1735–88), a notable botanist who accompanied Pennant on the Scottish tour of 1772 and who prepared *Flora Scotica* (1778), extended over a decade or more and was furthered by Lightfoot holding the living of Shalden (north of Alton) from 1765 to 1777.

John White . . . published: John White died in 1780 and W notes (NJ 29/30 May 1785) that the MS of his brother's work on Gibraltar was then at Selborne in his hands; the introductory section of this work was not published until 1911, and the remainder of John White's endeavours (except for part of his 'Flora') appears no longer to be extant.

223   *relieved my vine . . . annoyance*: owing to the endeavours of Thomas
      Hoar, W's manservant, NJ recording, for example, on 31 May
      1785: 'Thomas persists in picking the *cocci* off the vine; & has
      destroyed hundreds.' Hoar shared his master's interest in natu-
      ral history and kept records of the weather and other pertinent
      events when W was away from Selborne: he died in 1797, aged
      83.

224   *Not long since . . . limits*: the main details and much of the
      phrasing of what follows are entered in NJ for 26/27 Oct. 1782,
      at which time W was staying with his brother Henry, at Fyfield.

225   *Gold . . . stews*: goldfish were first introduced to England in the
      1690s, but general awareness occurred only after 1728, when a
      large number were presented to Sir Mathew Dekker, who
      distributed them around London. By 1769 Pennant was of the
      opinion that they were 'quite naturalized in this country, and
      breed as freely in the open waters as the common carp' (*BZ* iii).

226   *at times . . . November*: this general statement occurs in NJ at 10/
      11 Nov. 1781, a flight in November having been seen in 1777 (4
      Nov.) and in 1779 (5 Nov.).

      for several evenings . . . hill: this was on 13 and 14 Oct. 1780 (see
      NJ).

228   *There are three creatures . . . distance*: except for the first sentence,
      this whole para. occurs in NJ in late June 1787.

229   *why harmony . . . over*: the music was occasioned by the visit to
      Selborne (6 Sept. to 14 Oct. 1782) of relatives from Rutland;
      during the visit, W's nieces, Mary and Elizabeth Barker of
      Lyndon Hall (aged 22 and 18 respectively), played 'elegantly on
      the harpsichord . . . day after day with very lovely lessons from
      Niccolai, Giordani, and several other modern masters, in a very
      agreeable manner' (W letter to Ralph Churton, 4 Jan. 1783). So
      apposite to his experience was the passage that follows that W
      incorporated it in his writings on several occasions, including a
      letter (of 22 Jan. 1783) to Mary Barker, and was profuse in his
      thanks to the supplier of the same, another niece, Mary (Molly)
      White.

      Peireskii: Nicolas-Claude F. de Peiresc (1580–1637), a Provençal
      naturalist, whose biography was written by his friend Pierre
      Gassendi (1592–1655), French philosopher, mathematician, and
      mentor to Molière.

230   *This bird . . . walks*: the bird was a lesser whitethroat, *Sylvia*

curruca; the account is almost verbatim from NJ (11/12 May 1782) although in that journal it *follows* the details given in the next para. of swallows at James Knight's pond.

he shot . . . wood-pigeon: W records this in NJ as occurring on 29 Nov. 1782.

232 *gentleman*: Charles Etty (son of Andrew Etty, vicar of Selborne 1758–84) was mate on several Indiamen. The Chinese dogs were brought home from his second voyage, in summer 1787; from his first voyage (1783–4, during which his ship was lost by fire off Ceylon), he had brought back humming-birds, ostrich and turtle eggs, fine shells, and live tortoises.

234 *stone-curlew . . . dark*: this observation is recorded in NJ at 1 Mar. 1788 and seems to be the last such observation W was able to include in *Selborne*; it post-dates by almost nine months the date W assigned to the completion of the volume (see end DB 66).

235 *the scriptures . . . him*: the Bible gives several references to the feeding of ravens. W must have in mind Psalm 147: 9. The 'little girl', as is indicated in marginalia to a White family copy of *Selborne* (at Selborne), was Mary White, daughter of W's brother Thomas.

In reading . . . remarks: W had read the *Observations* of Huxham (1692–1768) as early as 1782, at which time he noted the substance of this para. in NJ for 14/15 Sept.

236 *Newton Valence*: this was at the vicarage, the 'gentleman' being Edmund White (1755–1838) who became incumbent there in 1785 and whose experiments of the preceding para. W entered in NJ as occurring early in Sept. 1786. This whole letter, together with DB 59, was a late addition to *Selborne*, W writing to Molly, his niece in London, in Nov. 1787: 'Pray give the opposite letter [DB 60] to your husband, and desire him to insert it among those addressed to Mr Barrington, but before [those] which all concern the weather at Selborne. This will soon be followed by an other [DB 59], which I shall also extract from my Journals . . . Desire your father [Thomas White] to attend to the *barometer part* of this letter' (R. Holt-White, *Life and Letters of Gilbert White of Selborne* (London, 1901), ii. 171, 173).

237 *four following letters*: originally written as four letters; at a late editorial stage a long letter on winter weather was subdivided into three and the total became six.

a work . . . utility: a persuasive affirmation of W's intention that is overlooked by admirers of *Selborne* who value the text only as a pastoral idyll.

238 *north-walls*: W's knowledge of the benefits of a wall was derived mainly from his own 'north' wall, albeit erected for protection to fruit (see *Garden Kalendar*, July 1761). John Fothergill (1712–80), whose garden at Upton (near Stratford, London) W may have visited, was especially interested in tender species from across the world and his collection was rated by contemporaries as second only to that at Kew; he was also a notable philanthropist, taking an active part in the foundation of Ackworth (Yorkshire), a school for children from Quaker families, and made important contributions to the understanding of diphtheria and influenza.

it froze . . . nights: as observed from the contents of chamber pots, urine freezing at a temperature lower than fresh water.

239 *On the 3rd January . . . place*: the observational details for this period of cold weather are taken direct from NJ, the barometer hanging within W's own parlour.

243 *neighbour's laurel-hedge . . . vigorous*: as NJ shows, this hedge belonged to Richard Yalden, vicar at Newton Valence.

244 *when the weather . . . useless*: a good example of W's enthusiasm for extremes, since, at the time he discovered the ineffectiveness of the instrument, he also entered his irritation in NJ.

246 *produce of my garden . . . nests*: there is a useful caution here about reading *Selborne*: W's customary practice is to delete all personal (and sometimes, even, human) agencies from his writing. In this instance, however, *Selborne* adds to an earlier formulation, NJ at 15 Aug. 1783 merely reading: 'Took this morning by birdlime on the tips of hazel-twigs several hundred wasps that were devouring the goose-berries'.

247 *unlike . . . within the memory of man*: and therefore noted by many writers of the period including Parson Woodforde in his *Diary* and William Cowper in *The Task*. The cause was the eruption of Skaptär-jokull, Iceland. The dates W gives for the effects echo the period of the volcano's most intense activity.

249 *my neighbours' windows*: although NJ records 'many windows broken', W gives no indication of where in the village it occurred. Cf. note to 'nests' (p. 246) where, similarly, *Selborne* provides human information omitted in NJ.

250 *the bells . . . discharged*: this was at the home in South Lambeth of

W's brother Thomas; he had inherited a considerable estate after the death in 1776 of a distant relative and retired from his business in London as a wholesale merchant.

Mr Aiken . . . sort: *The Calendar of Nature* (2nd edn., 1785) by John Aiken (see Biographical Notes).

take . . . leave of you and natural history together: a formal literary conclusion only. Publication of *Selborne* initiated a new correspondence, with Robert Marsham (see Bell, 1877), and in 1792–3 W hoped that Barrington would present to the Royal Society a monograph, modelled on his earlier papers of 1774 and 1775, on the nightjar, *Caprimulgus europaeus*; and entries continued in NJ until Sat. 15 June 1793, a mere ten days before W's death on Wednesday 26 June.

BIOGRAPHICAL NOTES

AIKIN, John (1747–1822), doctor, author; educated at Warrington Academy (where his father was tutor); medical practitioner at Great Yarmouth (1784–92); friend of Priestley, Pennant, Erasmus Darwin, John Howard—whose biography he wrote, and Southey (the poet); brother of Anna Letitia (Mrs Barbauld).

Edited many works by English writers, including Samuel Butler, Dryden, Goldsmith, Milton, Pope, Spenser, Thomson; known to W through John White (when vicar at Blackburn), and published *Calendar of Nature* (2nd edn., 1785) which W referred to in his epilogue to *Selborne* (see end of DB 66). After W's death prepared *A Naturalist's Calendar* (1795)—observations on natural history extracted from W's journals additional to those incorporated in *Selborne*; edited a second edition (2 vols.) of *Selborne* (1802).

BARKER, Thomas (1722–1809), scientist, meteorologist, theologian; grandson of William Whiston (Newton's successor as Lucasian professor of mathematics at Cambridge); contributed over twenty papers to Royal Society; published an important volume on *Comets* (1757), and three controversial theological works—on baptism (1771), on prophecy concerning the Messiah (1780), on demoniacks in the gospels (1783).

Maintained meteorological record at Lyndon, Rutland, for almost seventy years (a unique compilation for his period) and guided W in his observations of the weather. Married (1751) W's youngest sister, Anne, and visited Selborne on several occasions. Noted as a vegetarian from infancy, he drew W's praise when over 60 for riding 118 miles on horseback from Rutland to Selborne in three days, for having 'still a streight belly and [being] as agile as ever', and for running 'round Baker's hill [part of W's extended garden] in one minute and a quarter'; his son, Samuel, was W's most apt pupil in literary and natural history concerns.

BARRINGTON, Daines (1727–1800), lawyer, antiquary, and naturalist; sometime recorder at Bristol, and justice at Chester (1778–85); vice-president of the Society of Antiquaries, he was a friend of Dr Johnson and Boswell, and corresponded with Bishop Percy; admitted FRS 1767. Keenly interested in an understanding of the past and the improvement of the present, he showed in *Observations of the Statutes* (1766) how medieval law illuminated history; in 1767 he designed the 'Naturalist's Journal', sheets for the recording of observations, as a contribution to the discovery of a natural calendar; and included in *Miscellanies* (1781) a review of evidences for a circumnavigation of the globe via northern polar regions.

Sent W a copy of his 'Journal' format—at Pennant's suggestion; first met W in London in 1768 and encouraged him to pursue his observations; proposed in 1770 that W should prepare a fauna of Selborne; introduced W's monographs on the swallow, the martins, and the swift to meetings of the Royal Society (Feb. 1774; Mar. 1775); annotated W's 'Naturalist's Journals' for 1774 and 1775; and invited W, in 1775–6, to collaborate with him on a volume of natural history.

DERHAM, William (1657–1735), clergyman, natural historian; vicar at Upminster, Essex, from 1689; elected FRS 1702. Published a variety of papers in *Philosophical Transactions*, on the barometer, the 1703 Great Storm, wasps, the will o' the wisp, for example, and delivered (1711, 1712) the Boyle Lectures, which appeared the following year under the title *Physico-Theology, or a Demonstration of the Being and Attributes of God from his Works of Creation* (1713); this volume became a well-known text (it was translated into French, German, and Swedish); and a twelfth edition appeared in 1754 and was followed by *Astro-Theology* (1715), a similar demonstration 'from a Survey of the Heavens'.

Sometime chaplain to the Prince of Wales (the future George II), Derham was installed canon of Windsor (1716), edited works of John Ray, and made a large collection of birds and insects; one of W's intellectual mentors.

HALES, Stephen (1677–1761), clergyman, scientist, humanitarian; minister at Teddington, adjacent to Hampton Court, from 1709, and rector at Farringdon, neighbouring Selborne, from 1722; FRS 1718, member of French Academy 1753. Active in court circles (he was offered, but declined, a canonry at Windsor), his principal work was as a scientist and, more particularly, as a physiologist—showed in *Vegetable Statics* (1727) that sap in plants does not circulate in a fashion analogous to blood in mammals; work on artificial ventilation improved conditions in granaries, ships, and prisons.

Mocked for his experimental zeal; a contemporary poet wrote:

> Green Teddington's serene retreat
> For Philosophic Studies meet,
> Where the good Pastor Stephen Hales,
> Weighed moisture in a pair of scales,
> To lingering death put mares and dogs,

And stripped the skins from living frogs.
Nature he loved, her Works intent,
To search or sometimes to torment.

He was a founding member of the Society for the Encouragement of
Arts and Manufactures and Commerce (later, the Royal Society of
Arts), and admired by W, who wrote about his 'benevolent and useful
pursuits', and of a mind 'replete with experiment'; he is an exemplary
figure of one strand of the century's life—a man concerned to promote
human well-being and to find solutions to some of the practical
difficulties that beset daily living.

LINNAEUS, Carl (1707–78), Swedish naturalist; studied medicine at
Lund and Uppsala; journeyed to Lapland (1732), Holland (1735–8),
where he qualified as a doctor; practised as a physician in Stockholm
(1738–41); professor of medicine and botany at Uppsala (1741–72).

  Published many works, including *Systema Naturae* (1735), *Hortus
Cliffortianus* (1738), *Flora Suecica* (1745), *Fauna Suecica* (1746), *Species
Plantarum* (1753)—the starting-point for modern botanical nomenclature; *Systema Naturae* (vol. i, 10th edn., 1758)—the starting-point for
modern zoological nomenclature. Lasting memorial in principles of
binomial classification; applied from pharmacy the alchemist's signs
($\male$ Mars, iron) and ($\female$ Venus, copper) as symbols for male and
female; corresponded with naturalists throughout Europe; interested
in a country's customs and economy as well as its flora and fauna;
disciples travelled widely to collect specimens and report on modes of
life—for example, Kalm to America, Hasselquist to Egypt, Solander
and Sparrman (with Captain Cook) round the world.

  Advised John W (by correspondence) on some of his Gibraltar
specimens, and termed by W 'the greatest naturalist in Europe'—
albeit, in one respect, that of candour ('a readiness to acknowledge
mistakes on due conviction'), W gave the laurel to Ray.

MULSO, John (1721–91), clergyman; after office at Sunbury-on-
Thames, Thornhill (Wakefield), and Witney, Mulso was rector at
Meonstoke and a prebendary of Winchester.

  Friend of W from their days together at Oriel, Oxford, where they
shared extensive literary interests, and a somewhat florid correspondent with W throughout his life, he was the first recipient of W's poem
'The Invitation to Selborne', and visited Selborne with his family a
number of times, naming it as 'one of the most enchanting Spots in
England'. As a nephew of Bishop John Thomas (successively bishop
at Peterborough, Salisbury, and Winchester) and brother of Hester
(later Mrs Chapone), he linked W with both the periphery of royalty

and the literary circle of Samuel Richardson (with whom Hester entered into an extensive exchange of letters concerning his characterization of Clarissa, in the novel of that name). Encouraged W to publish and was of the opinion that *Selborne* would 'immortalize [his] Place of Abode as well as [himself]'.

PENNANT, Thomas (1726–98), traveller, writer on natural history; encouraged in his pursuits by the early gift of Willughby's *Ornithology* (1678) and by a visit to the Cornish naturalist and antiquary William Borlase. Toured extensively—Ireland (1754); the continent (1765), where he met Buffon, Voltaire, and Pallas; Scotland (1769–72); Isle of Man (1774); and made several journeys in England and Wales (1776, 1777, 1787). Correspondent of Linnaeus, he was elected to the Royal Society at Uppsala (1757); FRS 1767. Published *British Zoology* (1766; 1768–70), *Synopsis of Quadrupeds* (1771), *Arctic Zoology* (1784–7), *Indian Zoology* (1790), and accounts of his tours: some details in his works he obtained from informants (selected by the advertisement of 'Queries') and not only from his own personal observation.

First met W in London (1767); pleased to establish a reliable informant resident in central southern England (his own home being in Flintshire); assisted W and his brother with advice and books concerning creatures at Gibraltar and helped with identifications; cavalier over the use to which he put materials (specimens) he had been lent, but generous in his acknowledgement of the contribution made by W to his own work and incorporated W's corrections in a fourth edition of *British Zoology* (1776–7).

PLOT, Robert (1640–96), scholar, antiquary, writer on natural history; friend of Pepys and Evelyn; FRS 1677, and appointed secretary 1682; professor of chemistry at Oxford (1683); historiographer-royal (1688). Published *Enquiries* (1670) into natural phenomena (a form of advertisement for information, covering 'Heavens and Air', 'Waters', 'Earths', 'Stones', 'Metals', 'Plants', 'Husbandry', all within a particular region); *Natural History of Oxfordshire. Being an Essay towards the Natural History of England* (1677); *Natural History of Staffordshire* (1686).

RAY, John (1627–1705), naturalist, scholar; held several offices at Cambridge in 1650s; ordained (1660) but resigned fellowship at Trinity (1662) in face of Act of Uniformity; FRS 1667. Engaged in numerous, usually botanical, tours; Midlands and north Wales (1658), northern England and southern Scotland (1661), Wales and south-west England (1662), continental countries, including Germany and Italy (1663–6), south-west England (1667), northern England

(1668–1671); published *English Proverbs* (1670), *Wisdom of God* (1691) and many texts on aspects of natural history, including Willughby's *Ornithologia* (1676), *Methodus Plantarum Nova* (1682), *Historia Plantarum* (1704).

Although his professional work was published in Latin (thereby gaining widespread currency with international scholars), Ray is valued as the father of natural history in his native land or, as one historian puts it, 'the Aristotle of England'. Less dramatically and more modestly (in keeping with his own and with Ray's self-estimation), W described him as 'our countryman, the excellent Mr Ray, [who] is the only describer that conveys some precise idea in every term or word, maintaining his superiority over his followers and imitators in spite of the advantage of fresh discoveries and modern information' (*Selborne*, DB 10).

RÉAUMUR, Réné-Antoine Ferchault de (1683–1757), scientist, technologist, member of Royal Academy of Sciences (Paris) for almost fifty years; researched in several fields and actively prompted the systematic application of science to industry—his work leading to the foundation of steel-making, and of the tinplate industry; *Insectes* published 1734–42, and *The Art of Hatching and Bringing Up Domestick Fowls of All Kinds, at any Time of the Year—Either by means of the Heat of Hot-Beds, or that of Common Fire* (1750); devised Réaumur temperature scale, in which water boils at 80°C. Worked in the tradition of Hales, in which science and the useful are thought to be interdependent.

SCOPOLI, Giovanni Antonio (1723–88), doctor, naturalist. Professionally a medical practitioner, his *Annus I Historico-Naturalis* (1769), the first volume of a natural history of Carniola (formerly a province of Austria, now western Slovenia), provided W with the data that enabled him to identify an avian specimen from Gibraltar as a crag martin, *Hirundo rupestris*. Valued by W as a monographer, as someone who restricted his observations to a limited field in an age which, generally, attempted universal natural histories.

STILLINGFLEET, Benjamin (1702–71), naturalist, scholar; tutor at Felbrigg, Norfolk, for fourteen years to William Windham; toured continent (1737–43), and collaborated at Geneva in establishing 'a common room' devoted to theatrical performance in the natural manner; lived, on return to England, variously at Foxley, Herefordshire (home of Richard Price, father of Uvedale), Stratton-Strawless, Norfolk (home of Robert Marsham, correspondent of W from 1790 to 1793), and London, where he attended the literary evenings for conversation hosted by Mrs Eliz. Montagu and to which he (supposedly) wore 'blue stockings', thereby providing a sartorial soubriquet for women of intellect and purpose.

Proficient as a cellist, Stillingfleet wrote several dramas on biblical themes intended to be set as oratorios (*Paradise Lost* was performed at Covent Garden in 1760). *Miscellaneous Tracts relating to Natural History, Husbandry and Physick* (1759) introduced Linnaean principles to English readers and included his own 'Observations on Grasses'; to a second edition (1762) was added a 'Calendar of Flora', a comparison of natural events at Stratton and at Uppsala in the year 1755—this comparison led Stillingfleet to advocate the study of a natural calendar and resulted in W's determination to take up botany ('Flora Selborniensis—1766', also known as 'Calendar of Flora') and Barrington's devising of the 'Naturalist's Journal' (1767).

WHITE, John (1727–80), clergyman, naturalist, scholar of Corpus Christi, Oxford, he graduated in 1749 but was sent down the following year for being party to a marriage between 'a young Gentleman [of the College]' and an innkeeper's daughter from Wallingford, 'a House of no good Character', and for entertaining the bride and her sisters 'in his room at College'. Later, chaplain at Gibraltar (1756–72) and, from 1773, vicar at Blackburn, Lancashire; he became an apt pupil of W's interest in natural history and, in the period 1770–2, sent him several consignments of specimens. He corresponded with Linnaeus and prepared a natural history of Gibraltar, 'Fauna Calpensis' (*calpe*, Lat., pillar of Hercules). On his death his widow came to Selborne as W's housekeeper, and as such made no small contribution to the eventual appearance of *Selborne*.

WHITE, Thomas (1724–97), merchant, gentleman; member of the Apothecaries' Company; FRS 1777. By trade, partner in a wholesale ironmongery business in London, Thomas White inherited estates in 1776–7 from a distant relative (Holt) and retired to South Lambeth to pursue scholarly interests. Collected materials for a volume on the natural history and antiquities of Hampshire; contributed articles (including meteorological diaries, and a series on trees) to the *Gentleman's Magazine*.

Furnished the garden at Wakes (the family home at Selborne) with many specimen plants; experimented with growing American wheat; opened barrows on Selborne Down; and, together with his daughter, Mary (Molly), supported W in many of his enquiries, including the preparation of *Selborne*; reviewed his brother's book in the *Gentleman's Magazine*, 59 (1789). His 'Commonplace Book' is at Selborne Museum.

GLOSSARY

LANGUAGE usage in 1789 was very different from the usage of today. In *OED* there are 480 citations from White's writings, and recourse to a good dictionary (especially where a word meaning *seems* not to have changed) provides a valuable entry into the difference between his time and our time.

abroad: out of the house.

address: 'skill; dexterity' (Johnson)

admire: wonder; speculate

admits: records

affect: haunt, inhabit

aits: eyots; small islands, esp. those in rivers

aliment: nourishment; food

amusing, amusive (amuse): 'to fill with thoughts that engage the mind' (Johnson)

approximation: proximity; closeness

apterous: without wings

attrition: 'rubbing one against another' (Johnson); friction

aurelia: chrysalis of an insect

autopsia: ocular observations

awful: that which fills with reverence and/or fear

baiting place: feeding station

balk: the headland of a field (where the plough is unable to reach)

bark (TP 5): strip bark from timber (esp. oaks) to provide tan for curing hides

Bere: ancient forest in south-east Hampshire, north of Portsmouth

bloomy: fresh with the powdery deposit of ripeness

bottom: low ground; valley

callow: without feathers, hence naked, immature

campaign (TP 15): *see* champaign

canals: stretches of water, usually rectangular, in a garden

candour: 'sweetness of temper; purity of mind; openness' (Johnson)

cane: female weasel

cantoned: divided into small groups; quartered overnight

cardo: hinge of a bivalve

cart-way (TP 1): village thoroughfare

cere: membrane at the base of the beak in some birds in which the nostrils appear to be pierced

champaign, campaign: flat, open country; meadowland

characters: features, traits

chases: enclosed stretches of scrubland used for hunting

chesnut: chestnut (without the 't' before 1820)

churn-owl: nightjar

coddled: softened by heat

commerce: sexual intercourse

complacency: satisfaction; enjoyment; pleasure

concert: musical combination

congeners: members of the same genus or, today, because of reclassification, family

congeries: masses of small bodies heaped up together

constantly: regularly

conversation: general behaviour

correction: gaol

coverts: thickets, providing 'cover' for game

crabbed: difficult to understand

crumbs: food fragments, incl. meat and vegetable waste

cryptogame: breeding in secret

cryptogamia: flowerless plants such as ferns, lichens, algae

cunabula: cradles

curious: attentive, observant

deceiving: mistaking; bringing into error—but innocently, without ulterior motive

decent: good quality

declension: descent; lowering

decoy: a pond, with arms on several sides covered with netting, to which wild fowl are decoyed and there caught

defect (DB 8): absence; non-presence

denominations: species

desultory: 'roving from thing to thing ... immethodical' (Johnson); wavering; unsystematic (but not 'uncommitted')

discovers (DB 6): shows

diversion: capacity to avoid capture

dog-days: the hottest part of the year, part of July and August, named after Sirius, the dog-star

dor: any insect that flies with a loud humming noise as bumble-bee, hornet, cockchafer

drams: 'in weight the eighth part of an ounce' (Johnson)

dressed: cleaned and prepared for cooking

economy: management

eft: a general term for any newt-like lizard or lizard-like newt, Johnson citing an authority who claimed that the 'crocodile of Egypt is the lizard of Italy, and the *eft* in our country'

elegant: 'pleasing with minuter beauties' (Johnson); satisfying by its modesty and variety, rather than by grandeur and elevation

evolutions: wheelings and rollings in the air

exclude: hatch; give birth; lay (eggs)

faunists: writers on the animals of a particular country

fern: bracken

filices: ferns, pl. of *filix* (Lat.)

fledgling: a young fledged (feathered) bird, hence one that has just left the nest

flirt: 'quick, elastic motion' (Johnson)

forward (TP 1): conducive to early planting

freestone: stone of the Upper Greensand, able to be cut in any direction

furlongs: eighths of a mile (220 yards)

gage (gauge): unit of measurement

gallinae: typically birds such as domestic fowl that nest on the ground; from Lat. *gallus*, a cock

gibberish: 'the private language of . . . gipsies' (Johnson)

glance: rapid, oblique movement in the air

hanger: a wood on the side of a steep hill

haunt: 'to be much about any place' (Johnson); to frequent (with no overtones of 'threat')

hepatica: *Hepatica triloba* syn. *Anemone hepatica*, named thus because the three lobes of the leaves (*triloba*) have been compared to the three lobes of the liver (Gk. *hepar*)

hermitage: a mock hermit's cell used as a summerhouse

hinds: servants

hints: observations of behaviour

hoodwinked: blindfolded

hover: protect chicks with a downward sweep of the wing

hybernaculum: winter quarters of a hibernating animal

imbecility: 'weakness; feebleness of . . . body' (Johnson)

impertinent: irrelevant

impost stones: transoms; horizontal stone beams

infest (DB 25): are prevalent in

ingenious: talented; wise (with no sense of 'clever-clever')

insensible: imperceptible; without being felt or seen

irides: pl. of (Gk.) *iris*, coloured ring around the pupil

irregular: undulating and hilly

jealousy: vigilant solicitude

kenning: noticing

Lady-day: 25 Mar., the Feast of the Annunciation (the intimation of the incarnation of Jesus); quarter-day for payment of rent

lay-fields: land laid down as grassland, hence, pastures and meadows

leavened: transformed by expansion

legumens: plants producing seeds gathered by hand (e.g. peas and beans) rather than plants that are reaped (e.g. wheat, oats)

loaches: species of edible freshwater fish

lumping weight: good, generous weight

lurcher: collie–greyhound cross, used for catching rabbits and hares

magazines: storehouses

mast: fruit of forest trees—acorns, nuts

maws: stomachs

medicated: impregnated

meteor: weather phenomenon; signal, beacon (TP 24)

metheglin: 'drink made of honey boiled with water and fermented' (Johnson); mead, spiced or medicated

methodus: methodical (tabulated) arrangement of data—as given in DB 1, DB 2

Michaelmas-day: feast of St Michael, 29 Sept., a quarter-day

minuted: timed

monographers: those who write on one particular subject

mortification: vexation—without any sense of humiliation

mute: evacuate; void droppings

naked (TP 8): without aquatic vegetation and its associated organisms

nautili: species of cephalopod (tentacled), incl. those known as argonauts

nectarium: honey-sac

nice: 'accurate in judgment to minute exactness' (Johnson); discriminating; detailed

non-descript: not yet described and named in scientific literature

nostrum: 'medicine not yet made publick, but remaining in some single hand' (Johnson); family remedy

notorious: 'evident to the world; apparent' (Johnson); well-known

numbers (TP 44): verses, poetry

one with another: on average

ostensible: formally agreed

outlet: enclosure (garden, orchard, and so forth) attached to a house

parturition: 'state of being about to bring forth' (Johnson)

pectines: bivalves, shells of the Upper Greensand; pl. of *pecten* (Lat.)

peculiar: exclusive (TP 8); extraordinary (TP 17)

peremptorily: positively, as a matter of fact

phytologist: botanist

pitching: flooring

plastic: creative

pleasantly: amusingly; in the form of a pleasantry

pointers: dogs that point out game to sportsmen—usually by 'pointing' their bodies towards the game and quivering

policy: parkland surrounding a great house

polyglot: user of many languages (notes and songs); a bird that imitates the notes of other birds

preposterous: contrary to nature

provident: 'cautious; prudent with respect to futurity' (Johnson)

pulli: chicks or nestlings before they acquire true feathers (Lat. sing., *pullus*)

pulveratrices: birds that cleanse by dusting

purlieu: 'grounds on the borders of a forest' (Johnson)

quaint: scrupulously exact

quick-set: hawthorn, which grows quickly from live (quick) cuttings

radicals: roots of a word; words from which other words are derived

ranged: listed, arranged

rapine: plunder (of eggs and young birds)

reassume: undertake again—*not* continue after a break

reclaim: reform, or (for creatures) tame

recruit: rest, and replenish energies by feeding

remiges: main (quill) feathers in a bird's wing; pl. of Lat *remex*

retires: withdraws, migrates

retrices: tail feathers

sallets: salads

salutiferous: health-giving

sculk: 'lurk in places; to lie close' (Johnson); hide away

setting: hunting game with a dog trained to set, i.e. stand still and point its muzzle towards the hunted creature

sinks: open drains

some and some: 'now *some*, and then *some*' (Johnson, quoting Edmund Spenser)

sordid: niggardly, mean

spiculae: sharp-pointed crystals

squab: unfledged birds in the nest, esp. pigeons

stews: artificial ponds for fish (for the table)

straggle: 'be apart from . . . main body' (Johnson); reach (not necessarily 'behind')

stroll: venture

sublimes: raises to an elevated state; heightens

subulated: slender and tapering

succedaneum: substitute

supposititious: 'not genuine; put by a trick into the place . . . belonging to another' (Johnson)

suspicious: disposed to imagine something as likely

suture: jointed connection

team (TP 9): two or more horses or oxen to draw a cart

temperament: temperature

tenements: properties held by tenants

terrene: of the earth

train: brood

uropygium: rump

vagrants: wanderers

ventriloquous: from the stomach

vile (TP 17): low in station on the chain of being

vivary: any artificial enclosure for keeping animals, including a warren or a pond

warreners: keepers of warrens, ground in which rabbits were 'farmed'

wimble: 'instrument with which holes are bored' (Johnson); gimlet

winds (DB 61): 'power and act of respiration' (Johnson); breathing

withdraw: migrate

wither: edge of the shoulder-bone

yeoman-prickers: light horsemen

INDEX

Adams (old keeper) 21
Alice (Ayles) Holt 25, 27–8
Alton 11, 12, 13, 18–19, 20
Andalusia 38–9, 67, 78, 120, 123,
    216, 222
animals 41, 61, 88, 165–6, 181
Anne, Queen 21–2
aphide (smother-fly) 223–4, 246
ash, pollard 172–3
ass 40, 161
Astley, Mr 22
auks 86, 201

Banks, Mr 60
Barrington, Mr 48, 73
bats 33, 57
    common 33
    great 67, 81–2
bee-bird 121
beech 11–12, 85, 94, 123, 198, 235
bees 170–1, 184, 191, 238
Bell, Mr 167
Belon, Old 121
Bentley 28
bilberry, creeping 197
Bin (Bean's) pond 25, 197
birds:
    cleaning 116
    collections 73
    congregations 118–19, 126–7,
        165
    dispersion 194
    gender 114–15
    hard(thick)-billed 73, 110, 114
    locomotion 114, 199–201
    long-billed 111–12
    maternal affection 130–2
    migration discussed 121–3
    nidification 227–8
    pairing 71–2
    short-winged 38, 39, 45, 66,
        117, 141

soft-billed 39, 90, 91, 102, 110,
    114, 121
song 88, 104–8, 109, 111, 112,
    124, 201–4, 229
of summer passage 30, 36–7,
    101–2, 109, 113, 117, 121,
    141–2
of winter passage 52, 66, 85,
    103, 123
black-cap 30, 33, 44, 77, 89, 102,
    105, 109, 111, 113, 117, 141
blackbird 42, 65, 88, 105, 107,
    111, 119, 243
blind worm 47
bogs 20, 25, 234
botany:
    local plants 196–9
    worthiness of 194–6
brambling, great 66
Buffon, De 66
bullfinch 41, 86, 107
bull's head (miller's thumb) 33
bunting 38, 107, 117
bustard 117
butcher-bird:
    great ash-coloured 84
    red-backed 51
buzzard 45, 109, 142
    common 94
    honey- 94

canary 36
cane 41
castration 179–80
cat 72, 181, 245
cattle 26, 81, 165
chaffinch 37, 42, 86, 106, 110, 118,
    119
chicken 130, 166, 203–4
chough, Cornish 85, 117
Coccus vitis viniferae 221–3
cockchafer 60, 83

colemouse 91
Cressi-hall 56, 58
cricket:
  field- 209–11
  house- 212–13
  mole- 210, 213–15
crocus 198–9
cross-beak 32, 103, 117
cross-bill 32, 103, 142
crow 85, 200, 201, 241
  Royston-(grey) 103, 123
cuckoo 44, 102, 110–11, 112, 116, 123, 176–7
cuckoo-pint (arum) 41
Cumberland, Duke of 22

deer 22, 23, 28, 165
  fallow 28, 40
  moose- 69–71, 76
  red 21–3, 28
Derham, Mr 30, 45, 61, 79
dogs 232–3
domestic fowl 202–3
doves 201
  house- 95, 96
  ring- 95, 97, 107, 110, 123, 162, 200
  stock- 85, 95, 96, 103, 123
  turtle 44, 102
ducks:
  domestic 203
  wild 21, 25, 33, 45, 86, 103, 122, 132, 201, 202

earth-worm 182–3
echoes 189–93, 235–6
Ekmarck the Swede 125
elder, dwarf 198
Ellis, Mr John, FRS 47
elm, broad-leaved 14
Elmer, Mr 27

falcon (falco) 31–2, 34
  peregrine (haggard) 230–1
fieldfare 38, 63, 65, 69, 103, 119, 123
filices 19, 196

fish 24, 33, 45, 48–9, 54–5, 57, 89
  gold (silver) 224–5
flamingo 216
fly-catcher (stoparola) 30, 33, 44, 71, 90, 102, 113, 117, 130
fossils 15–16, 21, 234
Fothergill, Dr 239
frogs 45–6, 47
fungi 196

geese 103, 201, 202, 203
gentian, autumnal 197
Gibraltar 76–7, 130, 222, 230
goat-sucker (churn-owl, fern owl) 57, 58, 89, 102, 111, 177, 200, 201
goldfinch 36, 105, 111
grasses 195–6
greenfinch 36, 105, 200
greyhen 21
gross-beak 32, 103, 117
gross-bill 32, 103, 142
gypsies 166–7

Hales, Dr 175–6
Hanger, The 11, 12, 16, 19, 94, 184, 197, 198
hare 19, 66, 181, 241
Hasselquist 38, 216
hawks 121, 161, 202
  ring-tail 177
  sparrow- 87, 94, 130, 204
haws 35, 52, 60
heath fires 24
heath-cock (black game, grouse) 21
hedge-hog 67–9
heliotropes 205–6
hellebore
  green 197
  stinking 196–7
helleborine 198
hen-harrier 87, 199–200
herb Paris (one-berry, true-love) 197
heron 56, 58, 200
Herrisant, M 176, 177

hirundines 134, 141, 161, 202
 see also house-martin; sand-
  martin; swallow; swift
hogs 111, 180–1
Holt, The 25, 27–8
honeysuckle 246
hoopoe 32, 121
hops 13, 20, 191–2, 224
horticulture 186–8
house-martin (martlet) 33, 44, 71,
  95, 106, 193
 feeding 74–5
 hibernation 30, 36, 83, 185, 220,
  226–7
 migration 54, 59, 85, 88, 102,
  117, 161, 184–5
 monograph 135–9
Howe, Brigadier-General Emanuel
  Scroope, and Lady Ruperta
  27, 28
Huxham, Dr 235

iguana, mud 47
insects 39, 43, 57, 67, 73, 78–9,
  89, 116, 134–5, 156, 160,
  183, 184, 185, 221–2, 242
Ireland 48, 93

jackdaw (daw) 54, 55, 200
Johnson, Mr 44

kestrel (kestril, windhover) 87,
  130, 199
king-fisher 200
kite 121, 142, 199
Kramer, Mr 120

lacerta 47, 48
 see also newt
ladies' traces 197
land-rail 19, 44, 102
landslides 206–9
lapwing 21, 86, 127
lark 142, 201
 grasshopper- 43–4, 61, 71, 87,
  102, 106

sky- 105, 107, 116, 123, 201,
  241
 willow- 50
 wood- 88, 104, 105, 107, 110,
  201
lathyrus, narrow-leaved (wild)
  197
Legge, Mr and Mrs Henry Bilson
  27
leprosy 185–7
Lightfoot, Mr 222
Linnaeus, Carl 31, 37, 42, 47, 64,
  73, 79, 88, 102, 104, 115,
  117, 119–20, 125, 161,
  181–2, 201, 223, 224–5
linnet, common 37–8, 105, 111,
  117, 119
Lisle, Mr 180
live stock 23–4
lizards 52, 55, 58

magpie 200
maiden-hair, great golden 170
malm:
 black 12
 white 13, 14
martin
 bank-/sand-, *see* sand-(bank-)
  martin
 black-, *see* Swift
 house-, *see* house-martin
 migration 82–3
Martin, Mr Benjamin 239
maternal affection 181–2
Mazel, Mr Peter 49
Merret, Mr 47
Mezereon 198
Mordaunt, Mr 27
Motacilla trochilus 30
 see also willow-wren
mouse:
 field 131, 228
 harvest 30–1, 34–5, 39, 42
 house 39
 shrew- 66, 173

newt (eft) 47, 50
nightingale 44, 87, 102, 103, 106, 123, 124, 125, 141
nut-hatch 44, 228

oaks 13, 14–15, 21, 28
ophrys, birds' nest 197–8
osprey 84
otter 72
owl 45, 124, 133–4, 200, 201, 202
  barn 33
  brown 33, 134
  churn- (fern-), *see* goat-sucker
  eagle 66
  white 71–2, 132–3

parrots 200, 201
partridge 19, 21, 72, 116, 130, 243
peacock 80–1, 202–3
Pembroke, Lord 73
Pennant, Mr 114, 216, 230
periwinkle, less 197
petti-chaps 229–30
pheasant 19, 25, 110
pigeon 88, 200
Plestor, The 14, 172–3
Plot, Dr 20, 190
plover, stilt 216
Pochard 103
purple comarum (marsh cinque foil) 197

quail 19, 38
Queen's-bank 22

rabbit 23, 111
rat, water- 31, 66–7
raven 15, 107, 130, 201
Raven-tree, The 15
Ray, Mr 14, 30, 31, 35, 40, 42, 44, 61, 62–4, 66, 94, 115, 116, 126, 140, 188, 216
  reptiles 45, 46, 47, 48
red-breast 88, 91, 103, 105, 108, 109, 110, 111
red-wing 30, 103, 112, 119, 123, 125, 238

redstart 33, 44, 87, 89–90, 102, 106, 113, 117, 141
reguli non cristatus 33, 51
  see also willow-wren
reptiles 45, 45–8, 52, 117–18, 185
ring-ousel 35, 39, 52, 54, 55–6, 60, 74, 123
  location 85, 117
  migration 63, 65–6, 74, 83–4, 103, 142
rook 31, 41, 88, 107, 127, 142, 143, 200, 234–5, 241
rushes 168–70

St John's Wort (tutsan) 197
salicaria, species of 42, 60–1, 63, 66
sand-(bank-) martin 44, 102, 150–4, 161, 201
sandpiper 51
saxifrage, opposite golden 197
Scopoli, Dr 74–5, 76, 77, 114, 115, 120
Scotland 65–6, 93–4
sedge-bird 62–4, 86
Selborne:
  high wood 11, 97, 124, 196, 198
  manor 19
  parish 11, 19, 31, 90
  stream 13
  village 12, 19
  wells 13
sheep 24, 126, 141
sheep down (sheep-walk) 11, 12, 15, 60, 175
shell-snail 218
silk-tail, German 35, 104
silk-wood 170
snakes 45, 48, 64
snipe 21, 25, 30, 31, 33, 45, 86, 101, 103, 109, 200
snow-flake 41
soils 11, 12, 13, 16, 27–8
sparrow:
  hedge- 87–8, 91, 103, 105, 110, 112, 230

house- 71, 88, 116
reed- 66, 102, 104, 105, 111, 114
spiders, gossamer 163–5
squirrel 228
squnch (stonck) 64
starling 127, 200
Stawel, Lord 27, 28
Stillingfleet, Mr 30, 38
stilt, black-winged 217
stone-chatter 85, 103, 117
stone-curlew 42–3, 44, 45, 53, 64, 67, 77–8, 102, 234
stoparola, *see* fly-catcher
sundew:
    long-leaved 197
    round-leaved 197
Sussex Downs 12, 117, 140–1
swallow (house-swallow) 71, 95, 107, 113, 116, 122, 161, 193, 200, 202, 230
    feeding 75
    hibernation 29–30, 36, 121, 128, 185, 218
    migration 36, 38, 44, 58–9, 67, 82-3, 95, 101, 117, 161, 184
    monograph 143–8
    song 105, 111
    tail feathers 87
    Virgil on 140–9
Swammerdam, Mr 45
swan 87, 103
swift (black-martin) 33, 116, 125, 193–4, 201
    drinking 86
    feeding 67
    great Gibraltar 77
    hibernation 29
    migration 30, 67, 83, 85, 102, 220–1
    monograph 154–60

teal 21, 25, 26, 33, 86, 104, 132
teasel, small 197
Thames, the 13, 34, 36, 39, 150, 242
thrush 42, 65, 88, 119, 121, 130, 238, 242

missel- 35, 85, 106, 123, 161–2, 200
    song- 105, 107, 112
timber 13, 14, 20, 25, 28
titlark 85, 105, 107, 108, 109, 110, 116, 117
titmouse:
    blue 91, 92
    great 88, 91, 92, 106
    large 238–9
    long-tailed 91
    marsh 88, 91, 92, 106
toad 45, 49–50, 54, 55
tooth-wort 197
tor-ousel 74
    see also ring-ousel
tortoise 117–18, 129–30, 142–3, 184, 217–18, 219
trees, condensation 173–4
truffles 198

vegetation, weather's effect on 238, 239–40, 243, 245, 246
vine 44, 162, 221–3
viper (adder) 47, 178, 180

wagtail 86, 87, 91, 109, 110, 201
    grey 103
    white 38, 103
    yellow 38, 103
warbler, wood 30
wasp 246
water-ousel 117
weasel 87
weather 235–50
wheat-ear 38, 85, 92, 103, 117, 141
whin-chat 85, 92, 103, 117, 142
White, Mr Henry 48
White, Revd., John 222
white-throat 33, 44, 71, 89, 102, 105, 107, 110, 111, 113, 117, 141, 201, 230
whortle (bilberry) 197
widgeon (wigeon) 26, 33, 103
wild fowl 21, 26–7, 33, 103, 201

willow-wren 43, 44, 50–1, 88, 130,
  230
  largest 44, 71, 102, 106
  middle 44, 102, 106
  smallest 44, 101, 106
Willughby, Mr 39, 110, 154, 216
Wolmer forest 13, 17, 20–2, 28,
  125, 150, 197, 234
  boundaries 25
  lakes 26, 33, 86
  ponds 132
  road to 18–19
woodchat shrike 64
woodcock 19, 75, 103, 109, 119,
  120, 122–3, 123, 125, 126

woodpecker 44, 201
Woods, John 53–4
wren 88, 91, 103, 105, 107, 108,
  111, 117
  golden-crowned 45, 90–1, 103,
  106, 108
wryneck 43, 44, 90, 101
wych hazel (broad-leaved elm) 14

yellow monotropa (bird's nest)
  197
yellow-wort, perfoliated 197
yellowhammer 105, 108, 111
yew 35, 52, 60

THE WORLD'S CLASSICS

A Select List

HANS ANDERSEN: Fairy Tales
Translated by L. W. Kingsland
Introduction by Naomi Lewis
Illustrated by Vilhelm Pedersen and Lorenz Frølich

JANE AUSTEN: Emma
Edited by James Kinsley and David Lodge

Mansfield Park
Edited by James Kinsley and John Lucas

J. M. BARRIE: Peter Pan in Kensington Gardens & Peter and Wendy
Edited by Peter Hollindale

WILLIAM BECKFORD: Vathek
Edited by Roger Lonsdale

CHARLOTTE BRONTË: Jane Eyre
Edited by Margaret Smith

THOMAS CARLYLE: The French Revolution
Edited by K. J. Fielding and David Sorensen

LEWIS CARROLL: Alice's Adventures in Wonderland
and Through the Looking Glass
Edited by Roger Lancelyn Green
Illustrated by John Tenniel

MIGUEL DE CERVANTES: Don Quixote
Translated by Charles Jarvis
Edited by E. C. Riley

GEOFFREY CHAUCER: The Canterbury Tales
Translated by David Wright

ANTON CHEKHOV: The Russian Master and Other Stories
Translated by Ronald Hingley

JOSEPH CONRAD: Victory
Edited by John Batchelor
Introduction by Tony Tanner

DANTE ALIGHIERI: The Divine Comedy
Translated by C. H. Sisson
Edited by David Higgins

CHARLES DICKENS: Christmas Books
Edited by Ruth Glancy

FEDOR DOSTOEVSKY: Crime and Punishment
Translated by Jessie Coulson
Introduction by John Jones

The Idiot
Translated by Alan Myers
Introduction by W. J. Leatherbarrow

GEORGE ELIOT: Daniel Deronda
Edited by Graham Handley

ELIZABETH GASKELL: Cousin Phillis and Other Tales
Edited by Angus Easson

KENNETH GRAHAME: The Wind in the Willows
Edited by Peter Green

THOMAS HARDY: A Pair of Blue Eyes
Edited by Alan Manford

JAMES HOGG: The Private Memoirs and
Confessions of a Justified Sinner
Edited by John Carey

THOMAS HUGHES: Tom Brown's Schooldays
Edited by Andrew Sanders

HENRIK IBSEN: An Enemy of the People, The Wild Duck,
Rosmersholm
Edited and Translated by James McFarlane

HENRY JAMES: The Ambassadors
Edited by Christopher Butler

JOCELIN OF BRAKELOND:
Chronicle of the Abbey of Bury St. Edmunds
Translated by Diana Greenway and Jane Sayers

GWYN JONES (Transl.):
Eirik the Red and Other Icelandic Sagas

CHARLOTTE LENNOX: The Female Quixote
Edited by Margaret Dalziel
Introduction by Margaret Anne Doody

JACK LONDON: The Call of the Wild, White Fang, and other Stories
Edited by Earle Labor and Robert C. Leitz III

KATHERINE MANSFIELD: Selected Stories
Edited by D. M. Davin

KARL MARX AND FRIEDRICH ENGELS: The Communist Manifesto
Edited by David McLellan

HERMAN MELVILLE: The Confidence-Man
Edited by Tony Tanner

PROSPER MÉRIMÉE: Carmen and Other Stories
Translated by Nicholas Jotcham

MYTHS FROM MESOPOTAMIA
Translated and Edited by Stephanie Dalley

EDGAR ALLAN POE: Selected Tales
Edited by Julian Symons

PAUL SALZMAN (Ed.):
An Anthology of Elizabethan Prose Fiction

OLIVE SCHREINER: The Story of an African Farm
Edited by Joseph Bristow

TOBIAS SMOLLETT: The Expedition of Humphry Clinker
Edited by Lewis M. Knapp
Revised by Paul-Gabriel Boucé

ROBERT LOUIS STEVENSON: Kidnapped and Catriona
Edited by Emma Letley

The Strange Case of Dr. Jekyll and Mr. Hyde
and Weir of Hermiston
Edited by Emma Letley

BRAM STOKER: Dracula
Edited by A. N. Wilson

WILLIAM MAKEPEACE THACKERAY: Barry Lyndon
Edited by Andrew Sanders

LEO TOLSTOY: Anna Karenina
Translated by Louise and Aylmer Maude
Introduction by John Bayley

ANTHONY TROLLOPE: The American Senator
Edited by John Halperin

Dr. Wortle's School
Edited by John Halperin

Orley Farm
Edited by David Skilton

VIRGIL: The Aeneid
Translated by C. Day Lewis
Edited by Jasper Griffin

HORACE WALPOLE: The Castle of Otranto
Edited by W. S. Lewis

IZAAK WALTON and CHARLES COTTON:
The Compleat Angler
Edited by John Buxton
Introduction by John Buchan

OSCAR WILDE: Complete Shorter Fiction
Edited by Isobel Murray

The Picture of Dorian Gray
Edited by Isobel Murray

VIRGINIA WOOLF: Orlando
Edited by Rachel Bowlby

ÉMILE ZOLA:
The Attack on the Mill and other stories
Translated by Douglas Parmée

A complete list of Oxford Paperbacks, including The World's Classics, OPUS, Past Masters, Oxford Authors, Oxford Shakespeare, and Oxford Paperback Reference, is available in the UK from the Arts and Reference Publicity Department (BH), Oxford University Press, Walton Street, Oxford OX2 6DP.

In the USA, complete lists are available from the Paperbacks Marketing Manager, Oxford University Press, 200 Madison Avenue, New York, NY 10016.

Oxford Paperbacks are available from all good bookshops. In case of difficulty, customers in the UK can order direct from Oxford University Press Bookshop, Freepost, 116 High Street, Oxford, OX1 4BR, enclosing full payment. Please add 10 per cent of published price for postage and packing.